T0248858

Protein Phosphorylation

Protein Phosphorylation

Edited by **Michelle McGuire**

New York

Published by Callisto Reference,
106 Park Avenue, Suite 200,
New York, NY 10016, USA
www.callistoreference.com

Protein Phosphorylation
Edited by Michelle McGuire

International Standard Book Number: 978-1-63239-524-5 (Hardback)

Contents

Preface

This book compiles important information regarding protein phosphorylation and human health contributed by veteran scientists. The book elucidates the most significant research topics grouped under three sections: tyrosine kinases in receptor signaling and human diseases; protein kinases and phosphates in cell cycle regulation; and histidine kinases in two-component systems. It connects the basic protein phosphorylation channels with human health and diseases. This book also comprises of excellent figure illustrations and will be a valuable read for a broad spectrum of readers.

The information shared in this book is based on empirical researches made by veterans in this field of study. The elaborative information provided in this book will help the readers further their scope of knowledge leading to advancements in this field.

Finally, I would like to thank my fellow researchers who gave constructive feedback and my family members who supported me at every step of my research.

Editor

Tyrosine Protein Kinases
in Receptor Signaling and Diseases

Tyrosine Phosphorylation of the NMDA Receptor Following Cerebral Ischaemia

Dmytro Pavlov and John G. Mielke

Additional information is available at the end of the chapter

1. Introduction

Cerebral ischaemia (stroke) describes a condition wherein blood flow to the brain is reduced such that neurological function is disrupted, and neural cell death becomes possible. For several decades, stroke has remained a leading international cause of death and disability, which is the reason considerable effort has been applied to improve understanding of its pathogenesis; however, only a modest comprehension of the complex cellular processes underlying ischaemia-mediated cell death can currently be claimed. Our limited knowledge regarding how the brain is changed by an ischaemic event is part of the explanation for the absence of a successful clinical intervention, despite the examination of more than a thousand potential pharmacotherapies during the past fifty years [61, 69, 169].

Phosphorylation is the most broadly examined post-translational modification within the central nervous system [125, 222, 244]. Physiological shifts in neuronal activity, such as those that occur during memory formation, can lead to changes in protein phosphorylation; in a similar fashion, pathological changes in brain activity, such as those that occur during cerebral ischaemia, can also affect phosphorylation status. One principal means whereby the pattern of phosphorylation, especially at tyrosine residues [45, 80], can affect brain function is by regulating the activity of ionotropic receptors, which mediate the vast majority of rapid signal transmission. While the phosphoregulation of many ionotropic receptors has been examined, the NMDA sub-type of receptors that respond to the excitatory neurotransmitter glutamate has been the subject of a disproportionate level of attention due to its key role in neuronal communication.

To contribute to ongoing efforts directed at developing improved pharmacotherapies for stroke, the present review will provide a reflection on the manner in which ischaemic injury may alter neuronal physiology through changes in the tyrosine phosphorylation of the NMDA receptor; in particular, three goals will aim to be accomplished: (a) providing a general review of the primary upstream changes initiated by cerebral ischaemia, and, in so doing, highlighting the importance of the NMDA receptor (b) offering a summary of the structure and function of the NMDA receptor, and the evidence that establishes how the receptor's function and cellular distribution are altered by tyrosine phosphorylation (c) outlining what is known about how ischaemia may set in motion cellular changes leading to the aberrant, potentially harmful, and possibly self-amplifying over-activation of the NMDA receptor.

2. Cerebral ischaemia

2.1. Definition, prevalence, and risk factors

Insufficient cerebral blood supply may result from either the collapse of systemic circulation (leading to global ischaemia), or from the occlusion of a vessel that supplies a discrete region of the brain (leading to focal ischaemia). Although there are several possible causes of focal occlusions, they are predominantly the result of a foreign substance travelling within the cerebral circulation until the lumen becomes too narrow to permit further movement (embolism) [240]; the principal source of emboli is believed to be atherosclerotic plaques [187]. While uncontrolled bleeding from a vessel (haemorrhage) can also cause ischaemia of a focal nature, occlusion is thought to account for approximately 80% of focal events [198, 230].

For several decades, stroke has consistently been recognised as one of the leading causes of death worldwide, and one of the major causes of severe disability. Globally, over 15 million people per year are diagnosed with stroke, and a third of those afflicted die from complications relating to the injury [255]. In addition to significant medical consequences for affected individuals, cerebral ischaemia also presents enormous socioeconomic costs; for example, recent estimates place the direct and indirect annual costs associated with stroke in the United States at approximately 65 billion USD [47], while similarly constructed European estimates place the annual costs at approximately 77 billion USD [172]. Given that those who survive an ischaemic attack must cope with a variety of significant cognitive deficits (including aphasia, hemiparesis, and memory problems) that often lack treatment, the social costs of stroke are as enduring as they are significant.

Understanding the underlying causes of cerebral ischaemia requires an appreciation for the numerous genetic and environmental factors that contribute to its development, and that its determinants may be divided into non-modifiable and modifiable categories. The primary, and most significant, non-modifiable risk factor is age. The incidence of stroke rises exponentially with age, and the majority of strokes are seen within individuals who are older than 65 years of age [83, 203]. Gender is also an important consideration, for stroke incidence among men has consistently been shown to be approximately one-third greater than among women [203]. In addition, numerous American studies have indicated that the occurrence of stroke among multiple non-white demographic groups is greater than among white individuals, even when socioeconomic factors are considered [95, 202]. The principal modifiable risk factor for cerebral ischaemia is hypertension, and a large body of work has illustrated that the likelihood of stroke rises proportionately with increasing blood pressure [210, 256]. As well, cardiac disease, notably atrial fibrillation and coronary artery disease [203, 257], and metabolic disease, particularly type II diabetes and dyslipidaemia [189, 256], are also associated with elevated stroke risk. Finally, several lifestyle factors, including physical activity levels, cigarette smoking, alcohol consumption, and diet, have been shown to independently affect the potential for stroke development [13, 92, 93, 251].

2.2. Pathogenesis

Despite comprising only about 2% of total body weight, the brain receives 15% of cardiac output and consumes about 20% of the oxygen utilised by the body [28]. The brain's disproportionate circulatory demands are attributable to a high metabolic rate based almost exclusively upon cellular respiration; in addition, unlike most other organs, glucose stores in the brain are sufficient to cover energy requirements for only about one minute [83]. In a relatively quick manner, reduction of blood flow beyond a critical threshold results in the inability of neurones to fire action potentials, and, if sufficiently extensive, may lead to the failure of oxidative phosphorylation, which is the principal method of cellular energy production [5]. To avert the cellular energy crisis that rapidly follows reduced blood supply, cells in an affected area rely increasingly upon glycolysis; consequently, tissue concentrations of lactate and hydrogen ions increase dramatically, causing acidosis [214]. However, the comparatively meagre amount of energy provided by anaerobic metabolism provides limited compensation, and, in a short period of time, the lack of high-energy phosphate, combined with decreased pH, precipitates a multifactorial increase of membrane permeability.

A number of ionic gradients exist across the neuronal membrane (high intracellular $[K^+]$ and low intracellular $[Na^+]$, $[Cl^-]$, and $[Ca^{2+}]$), and these are quickly disrupted by the collapse of various energy-dependent pumps and transporters. Of particular note is Na^+/K^+-ATPase pump failure, which allows Na^+ to move into the cell causing neuronal depolarisation accompanied by the passive diffusion of Cl^- and water [126, 230]. In combination, the normalisation of ions across the cellular membrane and the concomitant movement of water lead to intracellular swelling that causes osmolysis (cytotoxic oedema), which significantly contributes to acute neuronal cell death [60].

Disrupted ionic homeostasis also leads to a dramatic and unregulated increase in the fusion of neurotransmitter storage vesicles with pre-synaptic membranes, which causes a massive release of vesicular content. Of the transmitters that flood the synapse following ischaemia, the most intensely studied has been the amino acid glutamate, which is the principal mediator of excitatory neurotransmission within the mammalian brain. The harm that might result from excessive glutamate was first observed in studies that found its systemic administration caused pronounced retinal degeneration [142, 173], a phenomenon described as "excitotoxicity". A substantial body of subsequent work has established that glutamate is a key element of neurodegeneration in general, and of ischaemic cell death in particular [119, 133, 199]. For example, glutamate efflux precedes widespread injury to cellular membranes and enzyme systems [2], the extracellular concentration of glutamate rises dramatically during ischaemia [63, 84], glutamate release is correlated with insult severity [29, 224], and glutamate receptor antagonists provide significant protection against ischaemic brain damage [119, 155, 215].

The widespread release of glutamate and the excessive stimulation of its high-affinity post-synaptic receptors are thought to act as critical elements that permit a profound rise of the intracellular calcium ion concentration. Calcium ions are involved in an array of neuronal functions, and their intracellular concentration is rigourously maintained at a level approximately 10^4 times lower than their extracellular concentration [143] by a combination of specialised binding proteins [8], sequestration within organelles [77], and extrusion [232]. One of the first studies to recognise the importance of calcium ions in cell death found that degeneration following axonal amputation occurred only when calcium ions were present in the bathing medium [208]. Subsequently, the essential role played by Ca^{2+} in glutamate-mediated cell death became established by studies that used mouse neocortical cultures [36, 88], rat hippocampal cultures [115, 190], and rat brain slices [57, 137]. Furthermore, work with culture [37, 67, 207], slice [276], and *in vivo* [15] models of ischaemia went on to reveal that a specific sub-type of glutamate receptor - the ionotropic NMDA receptor (section 3) - accounts for the majority of Ca^{2+} entry during and immediately after an ischaemic insult.

The dysregulation of intracellular Ca^{2+} has become recognised as a central branch point within the ischaemic cascade [11, 133, 213, 223], and serves as an important link between upstream activation of glutamate receptors and downstream stimulation of cell death mediators; for example, catalytic enzymes and free radicals. Several cytodestructive enzymes appear to be activated by cerebral ischaemia [119, 133, 185, 192], including proteases, phospholipases, and endonucleases. One set of enzymes that has received significant attention is the calpains, which are cytosolic cysteine proteases with variable Ca^{2+}-binding domains [217]. Calpains are ubiquitously expressed in the CNS, and a clear rise in their levels has been observed in models of both transient focal and global ischaemia [272, 280]. As well, activated calpains have been associated with damage to a variety of proteins [10, 241, 254], and calpain inhibitors have been found to provide a measure of protection in both culture [10] and *in vivo* models of ischaemia [12].

Free radicals have emerged as important players in the development of ischaemia-induced neuronal damage [3, 111, 133, 139]. The detection of free radical production following excitotoxicity caused by NMDA receptor stimulation has been clearly demonstrated in a variety of cultured rodent neurones [50, 76, 117, 194], and various groups have shown neuroprotection against excitotoxicity using antioxidant compounds [119]. As well, the mechanism of excitotoxicity-induced free radical production has been linked to Ca^{2+} by a report that demonstrated exposing isolated mitochondria to increasing calcium and sodium concentrations elevated free radical production [52], and another that showed removing extracellular calcium attenuated free radical production following NMDA application [50]. In addition to mitochondrial impairment, increased levels of reactive oxygen and nitrogen species are likely due to a combination of suppressed free radical scavengers and the elevation of formative enzymes, such as xanthine oxidase, cyclooxygenase, and nitrogen oxide synthases. Collectively, the cellular changes caused by increased free radical activity are extensive, and include lipid peroxidation, protein denaturation, and nucleic acid modification.

Slight changes in cerebral blood supply can be effectively managed by autoregulatory mechanisms that govern blood flow and oxygen extraction; however, decreases beyond this primary threshold initiate numerous cellular changes that become more severe in direct relation to the extent of the disturbance. The critical stages of stroke pathogenesis (figure 1) develop following a rapid and sustained drop of neuronal energy supply, and are generally thought to include a loss of ionic homeostasis, the unregulated release of the excitatory transmitter glutamate, the profound over-activation of glutamate receptors (particularly, the NMDA receptor), the dysregulation of intracellular Ca^{2+} levels, and the activation of a number of calcium-mediated internal changes that broadly affect cellular structure and function. While the exact manner and time course of ischaemia-mediated changes can be varied, and is influenced by factors such as insult severity, neuronal maturation, phenotype, and connectivity, the one thing held in common is the ultimate development of extensive neuronal cell death.

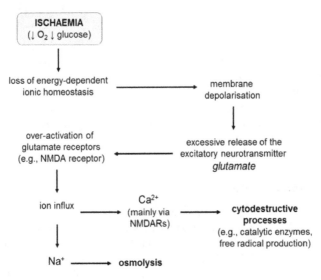

Figure 1. Summary of major elements in the early stages of ischaemic pathogenesis.

3. The NMDA receptor

3.1. Historical overview

Glutamate receptors (GluRs) mediate the majority of excitatory transmission in the vertebrate CNS, and participate in a number of physiological processes, including the formation of neuronal networks during development [43, 110], the pattern of ongoing synaptic communication [236], and the cellular plasticity believed to underlie learning and memory [21, 146]. In addition to an intimate involvement with the brain's physiology, glutamate responsive receptors are also of central importance in several neuropsychiatric conditions. For example, the GluRs have been implicated in neurodevelopmental disorders (e.g., schizophrenia) [59], mood disorders (e.g., depression) [149], chronic neurodegeneration (e.g., Alzheimer's disease, Parkinson's disease, and amyotrophic lateral sclerosis) [1, 22, 100], and pain transmission [20], in addition to brain injury (e.g., head trauma and stroke).

The broad influence of glutamate-mediated signalling upon synaptic function and dysfunction is attributable to the broad anatomical and cellular distribution of GluRs, and that they exist in two functionally and pharmacologically distinct varieties: metabotropic (mGluRs) and ionotropic (iGluRs). The metabotropic receptors are coupled to G-proteins, and, while structurally related to one another, do vary appreciably in their distribution and signal transduction mechanisms [175, 193]. The ionotropic receptors are non-specific cation channels that possess a common general structure, but vary considerably in both distribution and function [153, 175, 236]. As well, the iGluRs have been divided into three sub-types based upon relative selectivity to three exogenous agonists: N-methyl-D-aspartate (NMDA), α-amino-3-hydroxy-5-methyl-4-isoxazolepropionic acid (AMPA), and kainate.

Important preliminary evidence for diversity within excitatory neurotransmission was found in the early 1960s when the synthetic GluR agonist NMDA was shown to potently excite neurones [44]. Subsequent work in the 1970s, using radioligand binding and specific antagonists, established the existence of a specific NMDA subtype of iGluR (NMDAR) [154]. Following the advent of molecular cloning technology in the 1980s, a receptor complex possessing the functional characteristics ascribed to the NMDAR was characterised [158], which confirmed the existence of this particular iGluR sub-type, and helped to foment investigation into its physiopathological roles.

3.2. Subunit structure and assembly

The NMDA receptor is thought to be a heteromeric complex formed from a combination of four subunits: GluN1, GluN2 (with four known sub-types, labelled A-D), and GluN3 (with two identified sub-types, labelled A and B); notably, the nomenclature for GluR subunits has recently changed [42]. The GluN1 subunit has been shown to be essential to the formation of functional receptors [55], while the GluN2 and GluN3 subunits are believed to impart distinct gating and ion conductance properties [236]. Although the stoichiometry of subunits remains to be definitively resolved, endogenous NMDA receptors are thought to require the assembly of two GluN1 subunits with either two GluN2 subunits, or a combination of GluN2 and GluN3 subunits [19, 236]. Regardless of their ultimate arrangement, similar to other iGluRs, NMDARs are thought to be held within the endoplasmic reticulum until they assemble in a manner sufficient to permit counteraction of a retention signal [183].

One important limitation to improved understanding of NMDAR composition is the significant degree of developmental and anatomical heterogeneity that exists within subunit expression. The GluN1 subunit displays a peak degree of expression late in embryonic life before slightly declining to a relatively stable level of post-natal expression, while the GluN2 subunits vary considerably in their expression across the lifespan [53, 252]. For example, the GluN2A and GluN2C subunits are found post-natally, the GluN2B is expressed both before and after birth, although expression levels decline considerably between the early post-natal period and adulthood, and the GluN2D subunit is overwhelmingly restricted to embryonic development. As well, the GluN1 subunit is found in all central neurones, but a significant degree of anatomical heterogeneity exists among GluN2 subunits; in particular, the GluN2A and GluN2B subunits are found throughout the forebrain, the GluN2C subunit is limited to the cerebellum, and the GluN2D subunit is found predominantly within the midbrain [53, 156, 252, 253].

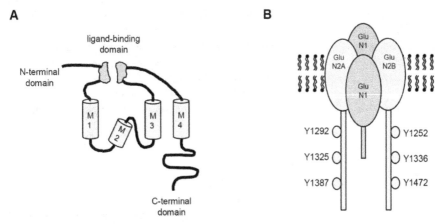

Figure 2. (A) Diagram of the structure for a typical NMDA receptor subunit. (B) Schematic illustrating one possible heterotetrameric combination of GluN subunits, along with identification of those GluN2 C-terminal tyrosine residues believed most relevant to phosphorylation-mediated changes in receptor gating and surface expression.

Despite being variably expressed, each NMDA receptor subunit shares a similar general architecture: a large extracellular region that consists of the amino-terminal and ligand binding domains, a pore-forming transmembrane region, and an intracellular region containing the carboxy-terminal domain [19, 56, 236] (figure 2). The N-terminal domain, at least in certain GluN2 subunits, is believed to allow receptor activity to be non-competitively inhibited by ligands such as zinc [177], although this may be an artifact of heterologous expression [261]. The adjacent ligand-binding domain is elegantly formed by two, non-contiguous segments that are separated by a portion of the polypeptide sequence thought to weave its way through most of the transmembrane region; as a result, conformational changes within the ligand-binding domain are thought to influence opening of the channel pore [56]. Four hydrophobic domains are believed to form the transmembrane region: the M1, M3, and M4 are predicted to cross the membrane as helices, while the M2, which lines the lumen of the pore, is expected to be a re-entrant loop that connects M1 and M3 [14, 19].

Among the NMDAR subunits, the C-terminal domain (CTD) is regarded as the most divergent region of the protein sequence [201], and can vary between 80-600 amino acids [56]. In addition to accounting for almost half the length of certain subunits (e.g., GluN2A and GluN2B), the CTD appears to be particularly important for intracellular signalling, trafficking, and localisation of the receptors due to the presence of multiple protein motifs that permit interaction with a variety of enzymes and scaffolding molecules. In particular, the intracellular region contains multiple locations for post-translational modifications, such as tyrosine phosphorylation [31, 125, 205, 236].

While the comparatively short CTD of the GluN1 does possess a tyrosine residue (Y837) [204], the subunit does not appear to experience tyrosine phosphorylation [121]; in contrast, each CTD of the GluN2 subunit contains 25 tyrosine residues, although not all of these residues will accept a phosphate group. On the GluN2A subunit, Y1292, Y1325, and Y1387 are thought to be the primary tyrosine residues subject to phosphoregulation [114]. On the GluN2B subunit, phosphorylation of Y1252, Y1336, and Y1472 has been reported [163]. Despite comprising a relatively small number of sites within the extensive CTD, tyrosine residues have become regarded as crucial points of convergence for signalling pathways that modulate NMDAR activity [170, 204, 205, 237].

3.3. Receptor function and cellular distribution

The basic pattern of excitatory signal transmission between the overwhelming majority of central neurones in the mammalian brain involves the pre-synaptic release of glutamate, its passage across the synaptic cleft, and its interaction with post-synaptically positioned GluRs. While basal synaptic transmission tends to be mediated by the AMPA sub-type of iGluR, periods of higher frequency synaptic activation (such as those that would tend to be present during an ischaemic event) recruit the NMDAR; the primary reason for the distinct activation profiles rests with a unique characteristic of the receptor. During basal transmission, the NMDAR's endogenous agonist (i.e., glutamate, which binds to an extracellular segment of GluN2 or GluN3 subunits) and its co-agonist (i.e., glycine, which binds to an extracellular segment of the GluN1 subunit) are present, yet the receptor remains functionally silent (i.e., ion conductance does not occur). The lack of basal NMDAR activity is attributable to a voltage-dependent blockade of the channel pore. At resting membrane potentials, external magnesium ions (which experience a significant inward driving force due to their high external concentration) enter the NMDAR pore, and bind in a manner that prevents further ion passage; however, membrane depolarisation of a sufficient magnitude and duration leads to the expulsion of Mg^{2+} from the pore, which permits the subsequent movement of cations [168].

Upon activation, glutamate receptors become permeable to both monovalent and, in some cases, divalent cations; however, the nature and degree of ion flux is not equivalent across iGluRs, and the NMDA sub-type has become acknowledged as the primary mediator of Ca^{2+} passage [57, 152, 236]. The ability to permit the entry of calcium ions is a primary reason that NMDARs make substantial contributions to both physiological and pathological phenomena [223, 236], although their pattern of cellular distribution also plays a role in their wide functional reach. Similar to many other ion channels, NMDARs are recognised as being widely dispersed across the cellular surface, however, information regarding the manner in which membrane bound receptors are distributed has only recently begun to be gathered. The available evidence indicates that NMDARs are not uniformly distributed at the plasma membrane, but can be divided into three spatially defined categories: synaptic, peri-synaptic and extra-synaptic [70].

The post-synaptic population exhibits the greatest density, relative to the other sub-regions, and responds directly to pre-synaptically released glutamate [70]. Peri-synaptic NMDARs have been operationally defined as those located within a 300 nm range of the post-synaptic terminal [70, 184, 275], and are believed to be activated by glutamate released from the pre-synaptic terminal following strong stimulation. Approximately half of all surface NMDARs, depending upon the developmental stage, are located extra-synaptically; however, relative to synaptic receptors, this compartment has a low density given the much broader area [72, 233]. Under physiological conditions, extra-synaptic NMDARs are unlikely to be activated by pre-synaptically released glutamate; however, they could be stimulated by glutamate released either from other sources [104, 128], or following the spillover of synaptic glutamate caused by ischaemia.

4. Regulation of NMDA receptors by tyrosine phosphorylation

4.1. Phosphorylation as a key determinant of ligand-gated ion channel function

Proteins exist as part of a complex system of elements that interact with one another to allow a cell to respond, directly or indirectly, to changes in its environment. Over several decades, the cellular distribution and molecular interactions of many proteins have been shown to be fundamentally regulated by post-translational modification in the form of phosphorylation [178, 238]. In essence, phosphorylation involves the protein kinase-mediated transfer of the γ-phosphate from adenosine triphosphate to a serine, threonine, or tyrosine residue of a substrate (although other amino acids may also be modified [40]), and is a process counteracted by protein phosphatases, which catalyse dephosphorylation.

Near the end of the 1980s, phosphorylation began to emerge as an important determinant in the function of a class of proteins essential for most neuronal communication – ligand-gated ion channels (LGICs). The first study to illustrate that LGICs could be regulated by phosphorylation revealed that NMDA currents recorded from cultured hippocampal neurones gradually declined during dialysis with an intracellular solution, but that the loss of receptor activity could be prevented by the addition of an ATP regenerating solution to the dialysate [148]. In the mid-1990s, further work not only confirmed that the NMDA receptor was indeed regulated by phosphorylation, but also that a critical site for the modification was at tyrosine residues [249]. Subsequent reports established that a variety of other LGICs, such as the inhibitory GABA-A receptor [159] and the nicotinic acetylcholine receptor [246], were also regulated by the activities of tyrosine kinases and phosphatases, and helped to establish the importance of phosphorylation in the control of synaptic function.

4.2. Regulation of NMDA receptors by tyrosine phosphorylation

Electrophysiological studies have established that NMDAR function is regulated by a balance between phosphorylation and dephosphorylation of tyrosine residues [31, 204, 205]. Within mammalian neurones, introducing an exogenous protein tyrosine kinase (PTK), or inhibiting endogenous protein tyrosine phosphatase (PTP) activity enhances NMDAR currents [249]; in contrast, inhibiting PTK activity [249, 250], or introducing an exogenous PTP [250] suppresses NMDAR currents. In addition, exogenous PTKs have been shown to potentiate currents mediated by recombinant NMDARs [32, 108]. During the past twenty years, members of the Src family of PTKs have emerged as the predominant tyrosine kinases responsible for mediating activity of the NMDA receptor.

4.2.1. Src Family Kinases (SFKs) and the brain: An introduction

The Src family of non-receptor, protein tyrosine kinases (SFKs) consists of several lower molecular weight proteins (52 to 62 kDa) that share a common domain organisation, which includes a catalytic region (the Src homology 1 or SH1 domain) and two regions that guide protein-protein interactions (the SH2 and SH3 domains) [170, 204]. Given that Src, the prototypical SFK, was initially identified as a proto-oncogene [218], SFKs in the brain were originally believed to influence only those processes related to the regulation of neuronal proliferation and differentiation [102, 116]. However, SFKs were subsequently shown to be expressed in differentiated, post-mitotic neurones, which suggested that these kinases might participate in neural activity past the point of early development [86, 220]. Notably, significant changes in neuronal plasticity and behaviour have been observed in adult mice lacking certain SFKs [68, 239], while the expression and activity of Src was shown to increase during spatial learning [278].

Within the mammalian nervous system, the expression of five members of the SFKs has been established; of these, Src [101], Fyn [221], and Yes [105] have been found within the post-synaptic density (PSD), which is a specialised region of the post-synaptic terminal thought to provide a molecular scaffold that helps regulate proper protein placement. In addition, Src [270], Fyn [267], and Yes [105] have been shown to be components of a large complex in the PSD that includes the NMDA receptor. Considered together, the biochemical analyses reveal that several SFKs have a spatial distribution appropriate to permit regulation of NMDAR function.

4.2.2. Evidence to support SFKs as modulators of NMDAR channel gating

When activated, ion channels, such as the NMDAR, passively conduct charged particles (i.e., they permit current flow) across the plasma membrane in a direction influenced by electrostatic forces and the ionic concentration gradient. The transition of a channel between closed and open states (i.e., from non-conducting to conducting) is referred to as gating, and requires a dramatic, albeit temporary, change in the channel's structure. The first evidence that SFKs might affect NMDAR function through alterations in channel gating was provided by a study that recorded channel currents from dissociated neurones [270]. A specific, high-affinity activator of SFKs [134] increased synaptic NMDAR-mediated currents, while an antibody that inhibited SFKs (anti-cst1; [195]), but not other protein tyrosine kinases, depressed NMDAR channel gating. A subsequent report, which recorded activity within neurones from acutely prepared brain slices, confirmed that the phosphopeptide SFK activator was able to significantly enhance NMDAR gating [141].

Additional support for the ability of SFKs to mediate NMDAR function was provided by a study that found protein tyrosine phosphatase alpha (PTPα) enhanced NMDAR-mediated synaptic currents in both cultured neurones and brain slices, while reducing PTPα activity with an inhibitory peptide reduced NMDAR currents [127]. Although the findings appear superficially paradoxical, PTPα is thought to activate SFKs by selectively dephosphorylating a residue in their regulatory domain, and thereby interfering with an intramolecular interaction that maintains the kinases in an inactive state [176]. In support of the proposed mechanism of action, PTPα has been shown to activate several different SFK members within cell lines [17, 87, 279], and SFK activity is substantially reduced in PTPα deficient mice [186]. As well, in cells lacking SFKs, or in which SFK activity is inhibited, PTPα has no effect on NMDAR currents [127].

While little doubt can exist that SFKs are involved in regulating NMDAR activity, the identities of the family members that might contribute to the process are not definitively known; however, considerable evidence has drawn attention to at least two kinases. The first candidate to be examined was Src, which was implicated by experiments illustrating that application of Src-specific inhibitors (the antibody anti-src1, and the peptide Src(40-58), which was the immunogen used to create the antibody) significantly decreased synaptic NMDAR-mediated currents and NMDAR channel gating [270]. As well, the Src-specific inhibitors were found to prevent the increased channel activity produced by application of a high-affinity SFK-activating peptide, which suggested that endogenous Src is required for SFK-mediated upregulation of NMDAR activity. To provide a structural complement to the functional studies, recent work with GluN2A subunits expressed in a heterologous expression system has found that Src directly interacts with segments in the C-terminus [75].

The second SFK member considered to have a role in NMDAR activity is Fyn, which was initially implicated by experiments wherein exogenous Fyn was shown to modulate glutamate-evoked currents mediated by recombinant NMDARs expressed in HEK293 cells [108]. Subsequently, co-expression of a constitutively active form of Fyn with GluN1 and GluN2B subunits in cerebellar granule cells was found to cause a significant increase in the amplitude of NMDA miniature excitatory post-synaptic currents [188]. In addition, several groups have reported that Fyn is able to phosphorylate both GluN2A and GluN2B within post-synaptic densities of the rodent forebrain [237].

Along with a direct effect upon NMDARs, SFKs may influence channel activity by phosphorylating proteins that associate with the receptor. For example, the post-synaptic density contains several proteins that are both tyrosine phosphorylated and potentially connected with NMDAR function, such as the scaffold protein PSD-93 [161, 206]. As well, phosphorylation of GluN2B may recruit signalling proteins, such as phosphatidylinositol 3-kinase, which has been shown to associate with the subunit [94, 182]. In addition, tyrosine phosphorylation of GluN2 subunits may prevent the loss of signalling molecules from the NMDA receptor by limiting degradation caused by calpain [18, 197], one of the principal proteases activated following ischaemia (section 2.2). Taken together, the variety of evidence suggests that SFKs may indirectly influence NMDAR function by altering the manner in which scaffolding and signalling proteins interact with GluN subunits.

4.2.3. Evidence to support SFKs as modulators of NMDAR trafficking

Rapid synaptic communication was traditionally believed to be altered by structural changes in ligand-gated ion channels that affected gating properties, such as mean open time and open probability. However, during the past fifteen years, cellular trafficking events, which modify the surface density of ion channels, have attracted considerable attention as an additional means whereby synaptic transmission can be regulated [41]. Although the inhibitory GABA-A receptor was the first LGIC found to undergo changes in cell surface expression in response to extracellular signals [245], the trafficking of other receptors, notably the NMDAR, has been the subject of growing interest over the past decade [70, 120, 183]. In particular, a number of studies have revealed that NMDARs are quite mobile, and undergo regulated trafficking between intracellular organelles and the plasma membrane [51, 64, 74, 124, 188, 191] and between extra-synaptic and synaptic sites [65, 71, 234].

The earliest evidence that NMDAR trafficking could be influenced through tyrosine phosphorylation was provided by a study wherein the repeated application of glutamate to heterologously expressed GluN1 and GluN2A subunits lead to an increase in receptor internalisation that could be prevented by application of Src [242]; using site-directed mutagenesis, the authors found that agonist-mediated dephosphorylation of a single tyrosine residue in the C-terminus of the GluN2A subunit was responsible for the reduced surface expression of functional channels. Subsequent studies wherein tyrosine phosphatase activity was reduced have demonstrated that phosphorylation of the NMDAR is positively associated with its surface expression. For example, brief treatment of cultured striatal neurones with a general inhibitor of tyrosine phosphatases increased the level of GluN2 subunit tyrosine phosphorylation and their surface expression [85]. As well, treatment of cultured cortical cells with a short interfering RNA to reduce the expression of a tyrosine phosphatase known to interact with the NMDAR lead to both an increase in its surface expression and the level of Ca^{2+} influx after agonist stimulation [24].

While enhanced surface expression of a receptor would reasonably be expected to increase its activity, the possibility that such a change will have a functional outcome is greater if the insertion occurs at synaptic membranes. In acutely prepared striatal slices from adult animals, a tyrosine phosphatase inhibitor was observed to increase the phosphorylation state of GluN2 subunits and increase their association with synaptosomal membranes; notably, treatment with a tyrosine kinase inhibitor had effects in the opposite direction [51]. An ensuing study that used tissue slices from the adult hippocampus and a separate set of tyrosine kinase and phosphatase inhibitors found the same general pattern tying together changes in GluN2 subunit phosphorylation status with synaptosomal membrane density [64]. Intriguingly, a further report revealed that the surface location to which GluN2 subunits are trafficked may be differentially affected by changes in tyrosine phosphorylation [65]. In particular, increasing the phosphorylation level of GluN2B subunits through the application of a tyrosine phosphatase inhibitor tended to increase the extra-synaptic expression of the NMDAR more greatly than its synaptic expression.

4.2.4. Balancing of SFK activity by tyrosine dephosphorylation

The level of NMDAR tyrosine phosphorylation is currently understood to be regulated by a balance between the activity of SFKs and protein tyrosine phosphatases [170, 204, 205] (figure 3). The PTPs are a large, structurally diverse family of enzymes [247] that play a number of important roles in the CNS, including contributing to the regulation of neural development [162, 219]. The striatal enriched phosphatase (STEP) is a brain-specific, non-receptor PTP that was originally found to be highly expressed within the adult rodent striatum [66, 138], and has subsequently been identified in other regions, such as the hippocampus [23]. While STEP immunoreactivity can be observed throughout the soma and processes of neurons [23], its presence at the post-synaptic density [174], a prominent structure at excitatory synapses, and its direct interaction with NMDARs [181], strongly suggest that the phosphatase contributes to signal transduction.

The ability of broad PTP inhibition to increase, and broad PTP activation to decrease, NMDAR gating in spinal cord neurones provided the initial evidence that the receptor was modulated by a phosphatase [250]. However, the identification of STEP as the putative candidate was not proposed until a report using the same experimental preparation showed that recombinant STEP reduced NMDAR activity, while the intracellular application of a function-blocking STEP antibody increased synaptic NMDAR-mediated currents [181]. Furthermore, within hippocampal slices, inhibition of STEP was shown to increase NMDAR activity, while recombinant STEP was able to occlude the development of a form of NMDAR-dependent synaptic plasticity.

The modulation of NMDA receptor function by STEP has become associated with its ability to affect the level of phosphorylation at a single tyrosine residue (Y1472) located in the distal portion of the GluN2B C-terminus [123, 196]. When dephosphorylated, the unique residue, which is part of a short motif (YXXφ, where X is any amino acid and φ = a bulky hydrophobic amino acid), promotes endocytosis by allowing the development of hydrophobic interactions between cargo molecules and clathrin adaptor proteins, such as AP-2 [235]. Notably, the phosphorylation of Y1472 and the association of GluN2B with adaptor proteins are inversely associated [164], and the overexpression of a Y1472 mutant unable to interact with AP-2 leads to a significant increase in the number of NMDA receptors at the synapse [188]. As well, STEP co-immunoprecipitates with the GluN1 subunit [181], directly dephosphorylates Y1472 [216], and, when deleted from the mouse genome, causes hyperphosphorylation of Y1472 in synaptosomal membranes prepared from the hippocampus [277]. Along with a direct effect upon Y1472, STEP may also act indirectly by dephosphorylating regulatory residues in the catalytic domains of Fyn, which is the principal SFK acting at Y1472 [165], and proline-rich tyrosine kinase 2 (Pyk2), which is an upstream activator of SFKs [262].

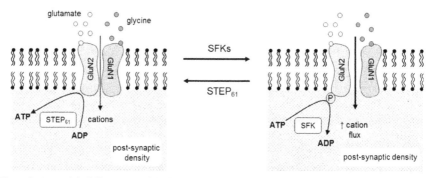

Figure 3. A simplified illustration of the key elements that regulate tyrosine phosphorylation of the NMDA receptor at the post-synaptic terminal.

5. Ischaemia-related changes in tyrosine phosphorylation of the NMDA receptor

5.1. Effects of ischaemia upon NMDA receptor subunit expression

Ischaemia causes a dramatic change in gene expression, which is largely attributable to a broad reduction in gene transcription and/or the inhibition of protein translation [109, 112]. While the general pattern of change suggests the level of most genes (and their protein products) will be reduced after ischaemia, several genes do show an increased level of expression; for example, certain ones associated with neuronal survival, such as members of the heat shock family of proteins. The first study to examine the effect of transient global ischaemia upon GluN2 subunits found that mRNA of both the GluN2A and GluN2B subunit declined in a progressive manner over 24 h of reperfusion, and that the transcriptional change was reflected in a pronounced loss of the proteins over the same time period [274]; notably, the study employed an antibody that detected an epitope shared between the subunits, so the relative nature of the protein loss was not apparent. A subsequent study that used subunit specific antibodies revealed that the pattern of reduced protein expression may not be equal, for the loss of the GluN2A subunit was greater [228]. In agreement with the possibility that GluN2 subunits may be differentially affected by ischaemic insult, hypoxia-ischaemia applied to rats at post-natal day seven caused a reduction of GluN2A protein levels within one hour of reperfusion, while a reduction in the GluN2B subunit was not observed; in contrast, similar experiments performed with animals at post-natal day 21 showed a significant reduction of only the GluN2B subunit [82].

While evidence does exist that the expression of GluN2 subunit proteins may be decreased by ischaemia, a comparatively greater amount of work suggests that the level of the subunits is not altered. For example, no change in GluN2B immunolabelling was detected in hippocampal synaptosomes prepared six hours after a period of transient global ischaemia in gerbils [179]. As well, the level of GluN2A and GluN2B subunit proteins in either hippocampal homogenates [227], or forebrain post-synaptic densities [226] prepared six hours following global ischaemia in rats did not appear to differ from the levels seen in control samples [227]; in agreement, a large group of studies that employed the same experimental approach failed to find an effect of the insult upon hippocampal protein expression of the GluN2A subunit (the GluN2B subunit was not examined) [33, 98, 136, 144, 248]. When the period of reperfusion was extended to 24 h, one study continued to find no change [97], while another found a slight, albeit significant, decline in GluN2A protein levels in whole hippocampal homogenates (again, the GluN2B subunit was not assessed) [135]. Collectively, the data strongly suggest that GluN2 subunit protein expression, particularly of the GluN2A variant, is not likely to be altered within the first 24 h after a brief period of global ischaemia in adult animals.

5.2. Ischaemia and the general cellular pattern of tyrosine phosphorylation

In the early 1990s, a brief stimulation of cultured hippocampal neurones with either glutamate, or NMDA was shown to cause a significant increase in tyrosine phosphorylation of mitogen activated protein kinase (MAPK, a class of kinases that respond to extracellular signals and mediate proliferation, differentiation, and cell survival) in a manner sensitive to blockade of the NMDA receptor [7]. Given the establishment of a connection between NMDA receptor activation and downstream tyrosine phosphorylation, subsequent reports sought to determine whether the excessive stimulation of the receptor that occurs during ischaemia would cause a similar pattern of change. Indeed, a brief period of bilateral carotid artery occlusion in gerbils (which would cause global ischaemia) was found to very quickly evoke a significant and transient increase in hippocampal MAPK phosphorylation that could be prevented through NMDAR antagonism [30, 107].

Once the association between ischaemia-mediated activation of the NMDA receptor and MAPK phosphorylation had been confirmed, a search for other proteins that might experience an induced change in tyrosine phosphorylation began in earnest. Within gerbils, global ischaemia was clearly shown to cause a rapid and significant increase in the level of tyrosine phosphorylation of a number of hippocampal proteins, particularly those with a higher molecular weight [171, 179, 269]. Within rats, both transient forebrain [99, 209] and global [34, 225, 228] ischaemia were found to induce a sustained increase in tyrosine phosphorylation of higher molecular weight proteins in the hippocampus. In addition, acutely prepared hippocampal slices subjected to *in vitro* models of ischaemia (i.e., oxygen-glucose deprivation) also displayed changes in the general level of tyrosine phosphorylation; however, while the magnitude of the effect was similar to that observed with the *in vivo* models, the direction of the effect was the opposite [6, 26].

Intriguingly, the post-ischaemic increase in tyrosine phosphorylation observed with *in vivo* injury did not appear uniformly throughout the hippocampus, which is composed of three major sub-fields: *cornu ammonis* (or CA) sector 1, CA3, and the dentate gyrus [54, 131]. Immunohistochemical labelling of hippocampal slices prepared at several time points following global ischaemia indicated that the CA3 and dentate gyrus appear to have the most intense initial increases in phosphorylation, but that, over several days, the CA1 region becomes more strongly labelled [269]. Immunoblotting revealed that the CA3 and dentate gyrus regions displayed greater tyrosine phosphorylation, particularly of higher molecular weight proteins, during the first two days after insult [228]. A disproportionate increase of tyrosine phosphorylation in the CA3-dentate gyrus was also observed in synaptosomes (sub-cellular fractions of pre-synaptic terminals that also include remnants of many post-synaptic sites) during the first day after ischaemia [99].

The elevated level of tyrosine phosphorylation routinely observed in various models of ischaemic brain injury would intuitively be attributable to increases in the activity of tyrosine kinases, decreases in the activity of tyrosine phosphatases, or a combination of the two. Within whole hippocampal homogenates, the level of Src was not found to change during a six hour period of reperfusion following either global ischaemia in adult animals [273], or hypoxia-ischaemia in pre-weanling animals [103]. In contrast, during the same length of time after global ischaemia the level of both Src and Fyn proteins were observed to experience about a twofold increase within the post-synaptic density [16, 34, 226].

Autophosphorylation of Y416 in the catalytic domain of SFKs is necessary to permit enzyme activity [204, 211, 266]. Within several hours of reperfusion following global ischaemia, a general increase in the level of tyrosine phosphorylation at hippocampal Src was observed [135, 136]; as well, a specific increase in Y416 phosphorylation was found within the whole hippocampus [38, 79, 248, 258, 273], and the CA1 [130, 264, 273] and CA3-dentate gyrus [79] regions. The level of phosphorylated Y416 was also found to be increased in both synaptosomes [160] and post-synaptic densities [16, 35, 160] prepared from the rodent forebrain region after the return of blood supply to the insulted area. In agreement with the immunoblotting data, *in vitro* enzyme activity assays confirmed that Src function in either whole hippocampal homogenates [78], or hippocampal synaptosomes [180] was clearly greater several hours after a brief period of global ischaemia.

While understanding how ischaemia may alter tyrosine kinase activity has been the principal focus of many studies, a developing body of work has chosen to examine changes at the level of tyrosine phosphatases; in particular, attention has been drawn to STEP. Glutamate-mediated excitotoxicity within cultured cortical neurones lead to the cleavage of STEP (i.e., STEP$_{61}$, a membrane-associated isoform localised to the endoplasmic reticulum and post-synaptic density) into a lower molecular weight isoform [66, 166, 263]. In agreement with the cell culture work, transient hypoxia-ischaemia in younger rats [81], brief focal ischaemia in adult rats [25], and global ischaemia in gerbils [89] caused the loss of STEP$_{61}$, and a concomitant rise of lower molecular weight STEP in the affected brain areas. Quite likely, STEP degradation is mediated by Ca^{2+}-activated proteases, such as calpain; for example, the *in vitro* incubation of isolated PSDs with calpain causes STEP$_{61}$ breakdown [81, 166], while treatment with a calpain inhibitor (calpeptin) prevents glutamate-activated STEP$_{61}$ cleavage [263]. Ischaemia also appears to reduce STEP levels at the transcriptional level by causing a rapid and significant reduction in its mRNA expression [25]. In addition, the exposure of cultured neurones to oxidative stress, using a free radical that undergoes increased production after ischaemia [4], results in greater oligomerisation of STEP$_{61}$, which causes a reduction in its activity level [46]. Considered together, the data indicate that ischaemia initiates a series of different processes that reduce STEP activity, and thereby contributes to enhanced levels of tyrosine phosphorylation.

5.3. The NMDAR as a specific substrate of ischaemia-mediated tyrosine phosphorylation

5.3.1. Evidence to illustrate ischaemic alteration of GluN2 phosphotyrosine status

The knowledge that ischaemia can bring about changes in the broad pattern of tyrosine phosphorylation, particularly of a glycoprotein with a molecular weight of 180 kDa, began to acquire additional significance after the identification of the GluN2B subunit as the major tyrosine phosphorylated, higher molecular weight glycoprotein associated with the post-synaptic density [157]. Shortly thereafter, came the first direct demonstration that ischaemia could specifically alter the tyrosine phosphorylation of NMDA receptor subunits; in particular, for several hours after a brief period of global ischaemia in adult rats, Takagi et al. [225] observed a rapid and dramatic rise in tyrosine phosphorylation of hippocampal GluN2A and GluN2B subunits. Although the modified phosphorylation pattern of the subunits was still clear 24 h after the insult, the GluN2A subunit consistently displayed a degree of change several times greater than that observed for the GluN2B subunit. As well, the magnitude of the effect was substantially greater in the hippocampus than either the cerebral cortex, or the striatum. The seminal report showed that ischaemia can cause a sharp and sustained rise in NMDAR tyrosine phosphorylation that differentially affects the receptor's constituent subunits and develops in an anatomically heterogeneous manner; in addition, the findings helped to influence a number of studies that confirmed ischaemia initiates cellular changes that ultimately modify the degree of tyrosine phosphorylation at GluN2 subunits (Table 1).

As noted, one of the first details to emerge regarding the pathological modification of NMDAR phosphorylation was the swift manner in which the change developed. Subsequent work revealed that within twenty minutes of reperfusion following a brief period of global ischaemia, a substantial rise in tyrosine phosphorylation of the GluN2A subunit could be detected [33, 34, 135]. As well, the magnitude of the GluN2A subunit modification was shown to increase during at least the first six hours of reperfusion, and was seen to slowly return to basal levels over 2-3 days [135, 228]. Using a similar *in vivo* model of ischaemia coupled with immunoprecipitation-based experiments, several groups confirmed a significant rise in the tyrosine phosphorylation of GluN2A subunits in either whole hippocampal homogenates [98, 136, 144, 226, 248], or CA1 homogenates [264] prepared within six hours of reperfusion.

While an increase in tyrosine phosphorylation of the GluN2B subunit was also detected after ischaemia, those studies that examined both subunits tended to find that the effect upon the GluN2A subunit was more pronounced [16, 160, 225, 226, 228]. One exception, was a study that applied global ischaemia to gerbils; however, the variation may have been, in part, attributable to a species difference [271]. In contrast to adult animals, ischaemic injury in weanling rats (post-natal day 21) was found to affect tyrosine levels of the GluN2A and GluN2B subunits in a similar fashion, and was shown to have a significantly greater effect upon the GluN2B subunit in pre-weanling rats (post-natal day 7); as well, the change in phosphorylation appeared to be more transient, for the levels had returned to baseline within a day of reperfusion [82, 103]. While the effect of oxygen-glucose deprivation (OGD) upon cultured neurones tended to match well with the effect of *in vivo* ischaemia upon the GluN2A subunit [248], the application of OGD to acutely prepared hippocampal slices caused an apparent decrease in the level of tyrosine phosphorylation at both subunits [6, 26]; notably, Src protein levels have been shown to rise in hippocampal slices after OGD [9], which suggests that the unexpectedly reduced level of phosphorylation might be attributable to a comparatively greater elevation in the expression and/or activity of tyrosine phosphatases.

In addition to an obvious effect upon GluN2 subunit phosphorylation at the cellular level (i.e., within whole homogenates of a brain region), several reports observed that ischaemia also clearly modified subunits located within synaptic compartments. For example, the immunoprecipitation of GluN2A and GluN2B subunits from post-synaptic densities after global ischaemia in the rat revealed a clear rise in tyrosine phosphorylation during the first six hours of reperfusion [34], while immunoblotting with antibodies directed against phosphotyrosine residues in the GluN2A (Y1387) and GluN2B (Y1472) subunits displayed a similar effect [16]. In agreement, several hours after transient restriction in blood supply to the rat cerebral cortex, tyrosine phosphorylation of both the GluN2A and GluN2B subunits was found to be increased in both post-synaptic densities and synaptosomes [160]. As well, the phosphotyrosine level of the GluN2B subunit in synaptosomes enriched from gerbil hippocampus was markedly increased within six hours of reperfusion after global ischaemia [179].

Reference	Species	Age or Weight	Ischaemia Model	GluN2A pY Change Relative to Control	GluN2B pY Change Relative to Control	Sample Size
16	rat	200-250 g	4VO & 6 h REP	pY1387: ~3.5 fold increase	pY1472: ~2.8 fold increase	5
26	rat	150-200 g	OGD	↓ in sample blot, no measurement	↓ in sample blot, no measurement	-
33	rat	250-300 g	4VO & 15 min REP	~4 fold increase	not examined	3
34	rat	200-250 g	4VO & 6 h REP	↑ in sample blot, no measurement	~4.0 fold increase	3
35	rat	200-250 g	4VO & 45 min REP	not examined	↑ pY1472 in sample blot, no measurement	1
82	rat	PND 7, PND 21	permanent right CCAO + hypoxia & 1 h REP	PND 7: similar PND 21: ~3.2 fold increase	PND 7: ~3.5 fold increase pY1472 (PND 7): ~2 fold increase PND 21: ~3.2 fold increase pY1472 (PND 21): ~8 fold increase	3-4
98	rat	250-300 g	4VO & 6 h REP	~6.2 increase	not examined	3
103	mouse	PND 7	permanent right CCAO + hypoxia	~2 fold increase (15 min REP)	~3 fold increase (15 min REP)	7
135	rat	200-250 g	4VO & 0-72 h REP	~6.8 fold increase (6 h REP)	not examined	3
136	rat	220-270 g	4VO & 6 h REP	~2.8 fold increase	not examined	4
144	rat	250-300 g	4VO & 6 h REP	~4.8 fold increase	not examined	3
160	rat	7-8 wks; 240-360 g	right MCAO & 4 h REP	~4.7 fold increase	pY1336: ~3.9 fold increase pY1472: ~3.3 fold increase	5
179	gerbil	60-75 g	BCAO & 6 h REP	not examined	~1.9 fold increase	3
225	rat	200-250 g	4VO & 6 h REP	~30 fold increase	~2.2 fold increase	5
226	rat	200-250 g	4VO & 1 h REP	~6.6 fold increase	~3.1 fold increase	4
228	rat	200-250 g	4VO & 1 h REP	CA1: ~13.5 fold increase CA3/DG: ~14.0 fold increase	CA1: ~3.0 fold increase CA3/DG: ~5.4 fold increase	5 5
248	rat	220-250 g	4VO & 6 h REP	~2.6 fold increase	not examined	4
264	rat	250-300 g	4VO & 6 h REP	~2 fold increase	pY1472: ~1.5 fold increase	5
268	rat	16 DIV	OGD	~5 fold increase	not examined	5
271	gerbil	50-70 g	BCAO & 3 h REP	~1.6 fold increase	~2.2 fold increase	6
273	rat	250-300 g	4VO & 6 h REP	~1.9 fold increase	not examined	4

Table 1. Studies examining GluN2 tyrosine phosphorylation following experimental ischaemia. 4VO, four vessel occlusion; BCAO, bilateral carotid artery occlusion; CCAO, common carotid artery occlusion; DIV, days *in vitro*; OGD, oxygen-glucose deprivation; PND, post-natal day; pY, phosphotyrosine; REP, reperfusion

5.3.2. Interaction of GluN2 with synaptic elements may influence its phosphorylation

Along with changes in the synaptic concentration and/or activity of those tyrosine kinases and phosphatases that act upon the NMDA receptor, the increased degree of tyrosine phosphorylation observed at GluN2 subunits after an ischaemic insult may also be attributable to changes in how the receptor directly interacts with SFKs. In particular, there may be greater association between phosphotyrosine sequences within GluN2 subunits and the segment of SFKs containing SH2 domains, which are relatively short amino acid modules believed to be necessary to permit the interactions that underlie many tyrosine based signalling cascades [150, 243]. Notably, ischaemia-mediated tyrosine phosphorylation was found to increase, by approximately twofold, the association of Src and Fyn with both GluN2A and GluN2B during reperfusion [226]. In agreement, a number of subsequent studies also observed greater association of SFKs with GluN2A during the first six hours following global ischaemia in adult rats [97, 98, 135, 248, 273]. As well, hypoxia-ischaemia applied to neonatal rats [103] and OGD of cultured neurones prepared from late-stage embryonic animals [268] also lead to greater co-immunoprecipitation of SFKs and the GluN2A subunit. Despite being examined to a lesser degree, the interaction between GluN2B and SFKs has also been clearly demonstrated after neonatal hypoxia-ischaemia [103], following global ischaemia in adult rats [179], and in response to global ischaemia within adult gerbils [271].

An additional component that may facilitate ischaemia-mediated changes in GluN2 phosphorylation is its enhanced interaction with a prominent post-synaptic scaffolding protein. The post-synaptic density 95 kDa (PSD-95) protein is a member of the membrane-associated guanylate kinase (MAGUK) family, and functions as an integral scaffolding protein in excitatory post-synaptic terminals [106, 265]. Like other MAGUK relatives, PSD-95 displays a modular structure that consists of three N-terminal PDZ domains (PDZ 1-3), an SH3 domain, and C-terminal guanylate-kinase domain. Through the PDZ1/2 domains, in particular, PSD-95 has been shown to bind with a conserved ES(E/D)V amino acid sequence located in the distal portion of GluN2 C-termini [113, 167], which permits PSD-95 to influence gating and surface expression of the receptor [39, 132]. A series of studies that employed multiple co-immunoprecipitation experiments helped to establish that an enhanced association of PSD-95 with the GluN2A subunit develops during the first few hours of post-insult reperfusion [33, 97, 144, 248, 273]. As well, an increased degree of binding with PSD-95 has been displayed by GluN2A subunits after OGD with cultured neurones [268] and by hippocampal GluN2B subunits within a few hours after global ischaemia in the gerbil [271].

Ischaemia clearly enhanced the interaction of GluN2 subunits with both SFKs and PSD-95, and increased the level of interaction between SFKs and PSD-95 [33, 49, 97, 144], presumably at its SH3 domain. As well, SFK-mediated tyrosine phosphorylation of GluN2 fusion proteins augmented their binding to PSD-95 [197], and co-expressing PSD-95 with the GluN2A subunit improved its ability to be phosphorylated by SFKs [231]. Considered together, these data suggest that the post-ischaemic increase of GluN2 tyrosine phosphorylation (section 5.3.1) is attributable to a pathologically heightened and, potentially, self-amplifying degree of interaction among GluN2 subunits, PSD-95, and SFKs. Our understanding of how the tripartite complex operates following ischaemia has been greatly aided by a set of elegant studies that used molecular level approaches to alter its assembly. Reducing by approximately one-third the hippocampal expression of PSD-95 protein, through the repeated intracerebroventricular injection of antisense oligonucleotides, sharply attenuated both the post-ischaemic rise in GluN2A tyrosine phosphorylation and the elevated binding of SFKs with the GluN2A subunit [98]. In addition, adenoviral-mediated overexpression of the PSD-95 PDZ1 domain was able to nearly eliminate ischaemia-mediated changes in GluN2A phosphorylation, and prevented the expected increased interactions between the GluN2A subunit, PSD-95, and Src [248].

5.4. Ischaemic changes that may contribute to increased SFK activation

The elevation of tyrosine phosphorylation within neurones, in general, and at GluN2 subunits, in particular, would seem to be a consistent feature in the array of changes that follows cerebral ischaemia. Proximally, the ischaemia-mediated interactions of a complex formed between GluN2 subunits, synaptic scaffolding proteins, and members of the SFKs provide at least one working explanation for the increased pattern of phosphorylation; however, the fashion in which these components begin to assemble is still unclear. A starting point in understanding, from a slightly more distal perspective, the reason for heightened GluN2 phosphotyrosine levels may be to focus on SFKs; specifically, to ask why might SFK activity rise after ischaemia? Fortunately, the answer to the question is beginning to take shape, and appears to involve the activity of two other, putatively interconnected, kinases. The first one is protein kinase C (PKC), which acts upon serine-threonine residues and exists in at least ten different isoforms, most of which are heterogeneously distributed within the brain [229]. While the various members of the PKC family have different mechanisms underlying their activation, many of them are stimulated by changes in Ca^{2+} dynamics and the generation of free radicals [27], which are key consequences following from the over-activation of NMDARs that characterises ischaemic injury (section 1). Also, the translocation of PKC to the post-synaptic density after ischaemia has been well established [35, 151].

The signalling cascade initiated by the stimulation of PKC is quite diverse, however, the ability of the activated enzyme to enhance NMDA receptor function [147, 260] and surface expression [118], is, to a large extent, attributable to its engagement of SFK activity. For example, the ability of PKC activators to potentiate NMDA-evoked currents in dissociated neurones was blocked by both tyrosine kinase inhibitors and Src-specific blocking peptides; as well, the PKC-dependent upregulation of the receptor was absent in neurones isolated from mice with a targeted deletion of the Src gene [140]. In addition, the stimulation of PKC activity in hippocampal slices was able to dramatically increase the level of phosphotyrosine detected within immunoprecipitated GluN2A and GluN2B subunits [73]. Direct support for the possibility that a PKC-dependent tyrosine kinase signalling cascade contributes to ischaemic changes in GluN2 tyrosine phosphorylation was provided by a study wherein a PKC inhibitor administered immediately after an ischaemic challenge was able to reduce, by approximately half, Src Y416 phosphorylation, general GluN2A tyrosine phosphorylation, and GluN2B Y1472 phosphorylation [35].

The second kinase that may serve to regulate post-ischaemic SFK activity is the proline-rich tyrosine kinase 2, which is a member of the focal adhesion kinase family and highly expressed within the brain [62, 259]. Through mechanisms that are still being uncovered, Pyk2 stimulation (permitted by phosphorylation of its Y402 residue) can be initiated through either depolarisation-induced Ca^{2+} influx, or PKC activation [101, 129, 212]; notably, the level of total and phosphorylated Pyk2 in the post-synaptic density increases sharply less than an hour after post-ischaemic reperfusion has begun [34, 35]. Upon phosphorylation, Pyk2 has been observed to interact with the Src SH2 domain to form a complex that enhances Src function [48, 101, 122]. In addition, the degree of interaction between Pyk2 and GluN2A is quickly amplified in the hippocampus during reperfusion after global ischaemia in either rat [135, 145], or gerbil [271]. Together, the data strongly suggest that ischaemia actuates Pyk2 and enhances its interaction with GluN2 subunits, which, in turn, likely allows Pyk2 to bind and activate the SFKs that would also have been drawn into a complex with NMDA receptors (figure 4).

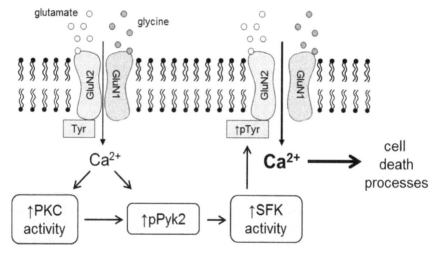

Figure 4. Model summarising the proximal steps leading to the ischemia-mediated increase in tyrosine phosphorylation of the NMDA receptor, and the possible association of its post-translational modification with upstream processes leading to cell death.

Direct support for Pyk2 as a critical component in the pathologic increase of GluN2 phosphorylation has been provided by a set of elegant studies that used either pharmacologic, or genetic approaches to downregulate the kinase's activity. The first study used lithium chloride (LiCl), which has been shown to protect neurones from ischaemic injury [200]. The application of LiCl for several days to cultured cortical neurones was able to dramatically reduce the basal level of both activated Src and phosphorylation at Y1472 of GluN2B [90, 91]; more importantly, administration of LiCl to animals for several days prior to global ischaemia was able to reduce insult-mediated increases in tyrosine phosphorylation of GluN2A [144]. As well, pre-ischaemic LiCl was observed to reduce both the phosphorylation of Y402 of Pyk2 and Y416 of Src, and to diminish the association of both Pyk2 and Src with GluN2A that generally accompanies ischaemic reperfusion [145]. The second study involved the intracerebroventricular injection of Pyk2 antisense oligonucleotides, which lead to a reduction of nearly one-quarter in the enzyme's protein expression within the hippocampus [136]. Animals in which Pyk2 levels had been downregulated displayed not only very little post-ischaemic change in the level of GluN2A tyrosine phosphorylation, but also very little change in the activity of Pyk2 and Src and their ability to be co-immunoprecipitated with GluN2A.

6. Conclusions

A short period of time after the circulation of blood to a region of the brain is substantially reduced, a series of escalating changes is initiated that can very quickly place neurones on a path leading to cell death. One of the principal elements in the pathologic cascade is the excessive stimulation of the excitatory NMDA receptor, which initiates a broad array of potentially cytodestructive changes; most germanely, a disruption in the usually strict regulation of the intracellular calcium ion concentration. One consequence of the elevated level of internal Ca^{2+} is the initiation of an enzyme cascade that involves either the direct, or indirect (through the serine-threonine PKC) stimulation of the tyrosine kinase Pyk2, followed by the subsequent activation of SFK members that go on to increase the tyrosine phosphorylation of GluN2 subunits. Notably, the post-translational modification of the GluN2 subunits is likely both a consequence, and, potentially, a cause of the post-ischaemic enhancement of the degree of interaction between the subunits and Pyk2, SFKs, and PSD-95. One important outcome of the ischaemia-mediated increase of GluN2 tyrosine phosphorylation that is quite likely concerns the enhancement of NMDA receptor gating properties and surface expression, which would serve as a form of signal amplification with the undesirable effect of contributing to increased cell death.

Preliminary support for the possibility that elevated NMDA receptor phosphorylation may contribute to post-ischaemic neuronal loss is found in a collection of studies that have sought to manipulate the cellular level of tyrosine phosphorylation for a neuroprotective effect. The first report in the area revealed that the intracerebral injection of specific inhibitors of protein tyrosine kinases minutes prior to a brief period of global ischaemia was able to prevent the significant degree of neuronal loss that would normally have been seen within the CA1 sub-field of the gerbil hippocampus a week after the insult [107]. Subsequently, the intraperitoneal injection of a selective inhibitor of Src shortly after forebrain ischaemia was found to significantly reduce not only the usual increases in Src activity and general tyrosine phosphorylation, but also the level of cell death observed in the hippocampus four days after the insult [171]. A group of successive studies observed that pretreatment of rats with PP2 (4-amino-5-(4-chlorophenyl)-7-(t-butyl)pyrazolo[3,4-d]pyrimidine), a selective SFK inhibitor, 0.5 h prior to global ischaemia was able to reduce by greater than half the usual increase in phosphorylated Src [130, 273] and GluN2A tyrosine phosphorylation [96, 273] seen after six hours of reperfusion; as well, cell density of the hippocampal CA1 sub-field five days after the insult was approximately 7-8 fold greater in the treated animals [96, 130, 273]. In addition, the injection of PP2 into neonatal mice shortly after hypoxia-ischaemia was able to reduce by about a third the degree of hippocampal cell loss observed several days after the insult [103].

By using tyrosine kinase inhibitors to control the level of GluN2 tyrosine phosphorylation after an ischaemic challenge, the excessive NMDA receptor activation that plays a critical upstream role in the resulting cell death may be, at least partially, addressed. As a result, a foundation may begin to be constructed that will serve as inspiration for the careful development of new approaches to address the discouraging absence of treatments for brain injury. Regardless of the outcome displayed by future studies that explore how understanding the causes and consequences of NMDA receptor tyrosine phosphorylation may be applied therapeutically, the activity will undoubtedly continue to add important details to our understanding of how the phosphorylation of brain proteins influences neuronal communication and synaptic dysfunction

Author details

Dmytro Pavlov and John G. Mielke*
School of Public Health and Health Systems, University of Waterloo, Waterloo, Canada

Acknowledgement

The preparation of this chapter was supported by a Discovery Grant from the Natural Sciences and Engineering Council of Canada (NSERC) awarded to J.G. Mielke.

7. References

[1]   Ahmed I, Bose SK, Pavese N, Ramlackhansingh A, Turkheimer F, Hotton G, Hammers A, Brooks DJ. (2011) Glutamate NMDA receptor dysregulation in Parkinson's disease with dyskinesias. Brain 134(4):979-986.
[2]   Ahmed SH, Hu CJ, Paczynski R, Hsu CY. (2001) Pathophysiology of ischemic injury. In Stroke Therapy, 2nd Edition, Butterworth-Heinemann, Boston, pp. 25-57.
[3]   Allen CL, Bayraktutan U. (2009) Oxidative stress and its role in the pathogenesis of ischaemic stroke. Int J Stroke 4(6):461-470.

---

* Corresponding Author

[4] Armogida M, Nisticò R, Mercuri NB. (2012) Therapeutic potential of targeting hydrogen peroxide metabolism in the treatment of brain ischaemia. Br J Pharmacol 166(4):1211-1224.

[5] Astrup J, Skovsted P, Gjerris F, Sørensen HR. (1981) Increase in extracellular potassium in the brain during circulatory arrest: effects of hypothermia, lidocaine, and thiopental. Anesthesiology (3):256-262.

[6] Au KN, Gurd JW. (1995) Tyrosine phosphorylation in a model of ischemia using the rat hippocampal slice: specific, long-term decrease in the tyrosine phosphorylation of the postsynaptic glycoprotein PSD-GP180. J Neurochem 65(4):1834-1841.

[7] Bading H, Greenberg ME. (1991) Stimulation of protein tyrosine phosphorylation by NMDA receptor activation. Science 253(5022):912-914.

[8] Baimbridge KG, Celio MR, Rogers JH. (1992) Calcium-binding proteins in the nervous system. Trends Neurosci 15:303-308.

[9] Barr CS, Dokas LA. (2001) Regulation of pp60(c-src) synthesis in rat hippocampal slices by in vitro ischemia and glucocorticoid administration. J Neurosci Res 65(4):340-345.

[10] Bano D, Young KW, Guerin CJ, Lefeuvre R, Rothwell NJ, Naldini L, Rizzuto R, Carafoli E, Nicotera P. (2005) Cleavage of the plasma membrane Na+/Ca2+ exchanger in excitotoxicity. Cell 120(2):275-285.

[11] Bano D, Nicotera P. (2007) Ca²⁺ signals and neuronal death in brain ischemia. Stroke 38(Suppl 2):674-676.

[12] Bartus RT, Baker KL, Heiser AD. (1994) Postischemic administration of AK275, a calpain inhibitor, provides substantial protection against focal ischemic brain damage. J Cereb Blood Flow Metab 14:537-544.

[13] Batty GD, Lee IM. (2002) Physical activity for preventing strokes. BMJ 325:350-351.

[14] Bennett JA, Dingledine R. (1995) Topology profile for a glutamate receptor: three transmembrane domains and a channel-lining reentrant membrane loop. Neuron 14(2):373-384.

[15] Benveniste H, Jorgensen MB, Diemer NH, Hansen AJ. (1988) Calcium accumulation by glutamate receptor activation is involved in hippocampal cell damage after ischemia. Acta Neurol Scand 78(6):529-536.

[16] Besshoh S, Bawa D, Teves L, Wallace MC, Gurd JW. (2005) Increased phosphorylation and redistribution of NMDA receptors between synaptic lipid rafts and post-synaptic densities following transient global ischemia in the rat brain. J Neurochem 93(1):186-194.

[17] Bhandari V, Lim KL, Pallen CJ. (1998) Physical and functional interactions between receptor-like protein-tyrosine phosphatase alpha and p59fyn. J Biol Chem 273(15):8691-8698.

[18] Bi R, Rong Y, Bernard A, Khrestchatisky M, Baudry M. (2000) Src-mediated tyrosine phosphorylation of NR2 subunits of N-methyl-D-aspartate receptors protects from calpain-mediated truncation of their C-terminal domains. J Biol Chem 275(34):26477-26483.

[19] Blanke ML, VanDongen A. (2009) Activation mechanisms of the NMDA receptor. In Biology of the NMDA Receptor, CRC Press, New York, pp. 283-312.

[20] Bleakman D, Alt A, Nisenbaum ES. (2006) Glutamate receptors and pain. Semin Cell Dev Biol 17(5):592-604.

[21] Bliss TV, Collingridge GL. (1993) A synaptic model of memory: long-term potentiation in the hippocampus. Nature 361(6407):31-39.

[22] Bogaert E, d'Ydewalle C, Van Den Bosch L. (2010) Amyotrophic lateral sclerosis and excitotoxicity: from pathological mechanism to therapeutic target. CNS Neurol Disord Drug Targets 9(3):297-304.

[23] Boulanger LM, Lombroso PJ, Raghunathan A, During MJ, Wahle P, Naegele JR. (1995) Cellular and molecular characterization of a brain-enriched protein tyrosine phosphatase. J Neurosci 15(2):1532-1544.

[24] Braithwaite SP, Adkisson M, Leung J, Nava A, Masterson B, Urfer R, Oksenberg D, Nikolich K. (2006) Regulation of NMDA receptor trafficking and function by striatal-enriched tyrosine phosphatase (STEP). Eur J Neurosci 23(11):2847-2856.

[25] Braithwaite SP, Xu J, Leung J, Urfer R, Nikolich K, Oksenberg D, Lombroso PJ, Shamloo M. (2008) Expression and function of striatal enriched protein tyrosine phosphatase is profoundly altered in cerebral ischemia. Eur J Neurosci 27(9):2444-2452.

[26] Braunton JL, Wong V, Wang W, Salter MW, Roder J, Liu M, Wang YT. (1998) Reduction of tyrosine kinase activity and protein tyrosine dephosphorylation by anoxic stimulation in vitro. Neuroscience 82(1):161-170.

[27] Bright R, Mochly-Rosen D. (2005) The role of protein kinase C in cerebral ischemic and reperfusion injury. Stroke 36(12):2781-2790.

[28] Brust JC. (2000) Circulation of the brain. In Principles of Neural Science, 4th Edition, McGraw-Hill, New York, pp. 1302-1316.

[29] Butcher SP, Bullock R, Graham DI, McCulloch J. (1990) Correlation between amino acid release and neuropathologic outcome in rat brain following middle cerebral artery occlusion. Stroke 21:1727-1733.

[30] Campos-González R, Kindy MS. (1992) Tyrosine phosphorylation of microtubule-associated protein kinase after transient ischemia in the gerbil brain. J Neurochem 59(5):1955-1958.

[31] Chen BS, Roche KW. (2007) Regulation of NMDA receptors by phosphorylation. Neuropharmacology 53(3):362-368.

[32] Chen C, Leonard J. (1996) Protein tyrosine kinase-mediated potentiation of currents from cloned NMDA receptors. J Neurochem 67:194-200.

[33] Chen M, Hou X, Zhang G. (2003) Tyrosine kinase and tyrosine phosphatase participate in regulation of interactions of NMDA receptor subunit 2A with Src and Fyn mediated by PSD-95 after transient brain ischemia. Neurosci Lett 339(1):29-32.

[34] Cheung HH, Takagi N, Teves L, Logan R, Wallace MC, Gurd JW. (2000) Altered association of protein tyrosine kinases with postsynaptic densities after transient cerebral ischemia in the rat brain. J Cereb Blood Flow Metab 20(3):505-512.

[35] Cheung HH, Teves L, Wallace MC, Gurd JW. (2003) Inhibition of protein kinase C reduces ischemia-induced tyrosine phosphorylation of the N-methyl-d-aspartate receptor. J Neurochem 86(6):1441-1449.

[36] Choi DW. (1985) Glutamate neurotoxicity in cortical cell culture is calcium dependent. Neurosci Lett 58(3):293-297.

[37] Choi DW. (1987) Ionic dependence of glutamate neurotoxicity. J Neurosci 7:369-379.

[38] Choi JS, Kim HY, Chung JW, Chun MH, Kim SY, Yoon SH, Lee MY. (2005) Activation of Src tyrosine kinase in microglia in the rat hippocampus following transient forebrain ischemia. Neurosci Lett 380(1-2):1-5.

[39] Chung HJ, Huang YH, Lau LF, Huganir RL. (2004) Regulation of the NMDA receptor complex and trafficking by activity-dependent phosphorylation of the NR2B subunit PDZ ligand. J Neurosci 24(45):10248-10259.

[40] Ciesla J, Fraczyk T, Rode W. (2011) Phosphorylation of basic amino acid residues in proteins: important but easily missed. Acta Biochim Pol 58(2):137-147.

[41] Collingridge G, Isaac J, Wang YT. (2004) Receptor trafficking and synaptic plasticity. Nat Rev Neurosci 5(12):952-962.

[42] Collingridge GL, Olsen RW, Peters J, Spedding M. (2009) A nomenclature for ligand-gated ion channels. Neuropharmacology 56(1):2-5.

[43] Constantine-Paton M. (1990) NMDA receptor as a mediator of activity-dependent synaptogenesis in the developing brain. Cold Spring Harbour Symp Quant Biol 55:431-443.

[44] Curtis D, Watkins J. (1961) Analogues of glutamic and gamma-amino-n-butyric acids having potent actions on mammalian neurones. Nature 191:1010-1011.

[45] Dabrowski A, Umemori H. (2011) Orchestrating the synaptic network by tyrosine phosphorylation signalling. J Biochem 149(6):641-653.

[46] Deb I, Poddar R, Paul S. (2011) Oxidative stress-induced oligomerization inhibits the activity of the non-receptor tyrosine phosphatase STEP61. J Neurochem 116(6):1097-1111.

[47] Demaerschalk BM, Hwang HM, Leung G. (2010) US cost burden of ischemic stroke: a systematic literature review. Am J Manag Care 16(7):525-533.

[48] Dikic I, Tokiwa G, Lev S, Courtneidge SA, Schlessinger J. (1996) A role for Pyk2 and Src in linking G-protein-coupled receptors with MAP kinase activation. Nature 383(6600):547-550.

[49] Du CP, Gao J, Tai JM, Liu Y, Qi J, Wang W, Hou XY. (2009) Increased tyrosine phosphorylation of PSD-95 by Src family kinases after brain ischaemia. Biochem J 417(1):277-285.

[50] Dugan LL, Sensi SL, Canzoniero LM, Handran SD, Rothman SM, Lin TS, Goldberg MP, Choi DW. (1995) Mitochondrial production of reactive oxygen species in cortical neurons following exposure to N-methyl-D-aspartate. J Neurosci 15(10):6377-6388.

[51] Dunah AW, Standaert DG. (2001) Dopamine D1 receptor-dependent trafficking of striatal NMDA glutamate receptors to the postsynaptic membrane. J Neurosci 21(15):5546-5558.

[52] Dykens JA. (1994) Isolated cerebral and cerebellar mitochondria produce free radicals when exposed to elevated CA^{2+} and Na^+: implications for neurodegeneration. J Neurochem 63:584-591.

[53] Ewald RC, Cline HT. (2009) NMDA receptors and brain development. In Biology of the NMDA Receptor, CRC Press, New York, pp. 1-15.

[54] Fanselow MS, Dong HW. (2010) Are the dorsal and ventral hippocampus functionally distinct structures? Neuron 65(1):7-19.

[55] Forrest D, Yuzaki M, Soares HD, Ng L, Luk DC, Sheng M, Stewart CL, Morgan JI, Connor JA, Curran T. (1994) Targeted disruption of NMDA receptor 1 gene abolishes NMDA response and results in neonatal death. Neuron 13(2):325-338.

[56] Frank R. (2011) Endogenous ion channel complexes: the NMDA receptor. Biochem Soc Trans 39:707-718.

[57] Garaschuk O, Schneggenburger R, Schirra C, Tempia F, Konnerth A. (1996) Fractional Ca2+ currents through somatic and dendritic glutamate receptor channels of rat hippocampal CA1 pyramidal neurones. J Physiol 491(3):757-772.

[58] Garthwaite G, Hojos F, Garthwaite J. (1986) Ionic requirements for neurotoxic effects of excitatory amino acid analogues in rat cerebellar slices. Neuroscience 18:437-447.

[59] Gilmour G, Dix S, Fellini L, Gastambide F, Plath N, Steckler T, Talpos J, Tricklebank M. (2012) NMDA receptors, cognition and schizophrenia - testing the validity of the NMDA receptor hypofunction hypothesis. Neuropharmacology 62(3):1401-1412.

[60] Ginsberg MD. (1996) Pathophysiology and mechanisms of ischemic brain injury. In New Approaches to the Treatment of Ischemic Stroke: Part I. Ischemic Stroke: A Disease Update. Excerpta Medica, New Jersey, pp. 11-17.

[61] Ginsberg MD. (2009) Current status of neuroprotection for cerebral ischemia: synoptic overview. Stroke 40(3 Suppl):S111-S114.

[62] Girault JA, Costa A, Derkinderen P, Studler JM, Toutant M. (1999) FAK and PYK2/CAKbeta in the nervous system: a link between neuronal activity, plasticity and survival? Trends Neurosci 22(6):257-263.

[63] Globus MY, Busto R, Martinez E, Valdes I, Dietrich WD, Ginsberg MD. (1991) Comparative effect of transient global ischemia on extracellular levels of glutamate, glycine, and gamma-aminobutyric acid in vulnerable and nonvulnerable brain regions in the rat. J Neurochem 57(2):470-478.

[64] Goebel SM, Alvestad RM, Coultrap SJ, Browning MD. (2005) Tyrosine phosphorylation of the N-methyl-D-aspartate receptor is enhanced in synaptic membrane fractions of the adult rat hippocampus. Brain Res Mol Brain Res 142(1):65-79.

[65] Goebel-Goody SM, Davies KD, Alvestad Linger RM, Freund RK, Browning MD. (2009) Phospho-regulation of synaptic and extrasynaptic N-methyl-d-aspartate receptors in adult hippocampal slices. Neuroscience 158(4):1446-1459.

[66] Goebel-Goody SM, Baum M, Paspalas CD, Fernandez SM, Carty NC, Kurup P, Lombroso PJ. (2012) Therapeutic implications for striatal-enriched protein tyrosine phosphatase (STEP) in neuropsychiatric disorders. Pharmacol Rev 64(1):65-87.

[67] Goldberg MP, Choi DW. (1993) Combined oxygen and glucose deprivation in cortical cell culture: calcium-dependent and calcium-independent mechanisms of neuronal injury. J Neurosci 13(8):3510-3524.

[68] Grant SG, O'Dell TJ, Karl KA, Stein PL, Soriano P, Kandel ER. (1992) Impaired long-term potentiation, spatial learning, and hippocampal development in fyn mutant mice. Science 258(5090):1903-1910.

[69] Green AR, Odergren T, Ashwood T. (2003) Animal models of stroke: do they have value for discovering neuroprotective agents? Trends Pharmacol Sci 24(8):402-408.

[70] Groc L, Bard L, Choquet D. (2009) Surface trafficking of N-Methyl-D-Aspartate receptors: physiological and pathological perspectives. Neuroscience 158:4-18.

[71] Groc L, Heine M, Cognet L, Brickley K, Stephenson FA, Lounis B, Choquet D. (2004) Differential activity-dependent regulation of the lateral mobilities of AMPA and NMDA receptors. Nat Neurosci 7(7):695-696.

[72] Groc L, Heine M, Cousins SL, Stephenson FA, Lounis B, Cognet L, Choquet D. (2006) NMDA receptor surface mobility depends on NR2A-2B subunits. Proc Natl Acad Sci USA 103:18769-18774.

[73] Grosshans DR, Browning MD. (2001) Protein kinase C activation induces tyrosine phosphorylation of the NR2A and NR2B subunits of the NMDA receptor. J Neurochem 76(3):737-744.

[74] Grosshans D, Clayton D, Coultrap S, Browning M. (2002) LTP leads to rapid surface expression of NMDA, but not AMPA, receptors. Nature Neurosci 5:27-33.

[75] Groveman BR, Xue S, Marin V, Xu J, Ali MK, Bienkiewicz EA, Yu XM. (2011) Roles of the SH2 and SH3 domains in the regulation of neuronal Src kinase functions. FEBS J 278:643-653.

[76] Gunasekar PG, Kanthasamy AG, Borowitz JL, Isom GE. (1995) NMDA receptor activation produces concurrent generation of nitric oxide and reactive oxygen species: implication for cell death. J Neurochem 65(5):2016-2021.

[77] Gunter TE, Pfeiffer D. (1990) Mechanisms by which mitochondria transport calcium. Am J Physiol 258:C755-C786.

[78] Guo J, Meng F, Zhang G, Zhang Q. (2003) Free radicals are involved in continuous activation of nonreceptor tyrosine protein kinase c-Src after ischemia/reperfusion in rat hippocampus. Neurosci Lett 345(2):101-104.

[79] Guo J, Wu HW, Hu G, Han X, De W, Sun YJ. (2006) Sustained activation of Src-family tyrosine kinases by ischemia: a potential mechanism mediating extracellular signal-regulated kinase cascades in hippocampal dentate gyrus. Neuroscience 143(3):827-836.

[80] Gurd JW. (1997) Protein tyrosine phosphorylation: implications for synaptic function. Neurochem Int 31(5):635-649.

[81] Gurd JW, Bissoon N, Nguyen TH, Lombroso PJ, Rider CC, Beesley PW, Vannucci SJ. (1999) Hypoxia-ischemia in perinatal rat brain induces the formation of a low molecular weight isoform of striatal enriched tyrosine phosphatase (STEP). J Neurochem 73(5):1990-1994.

[82] Gurd JW, Bissoon N, Beesley PW, Nakazawa T, Yamamoto T, Vannucci SJ. (2002) Differential effects of hypoxia-ischemia on subunit expression and tyrosine phosphorylation of the NMDA receptor in 7- and 21-day-old rats. J Neurochem 82(4):848-856.

[83] Hacke W, Hennerici M, Gelmers HJ, Kramer G. (1991) Epidemiology and classification of strokes. In Cerebral Ischaemia, Springer-Verlag, New York, pp. 31-52.

[84] Hagberg H, Lehmann A, Sandberg M, Nystrom B, Jacobson I, Hamberger A. (1985) Ischemia-induced shift of inhibitory and excitatory amino acids from intra to extracellular compartments. J Cereb Blood Flow Metab 5:925-929.

[85] Hallett PJ, Spoelgen R, Hyman BT, Standaert DG, Dunah AW. (2006) Dopamine D1 activation potentiates striatal NMDA receptors by tyrosine phosphorylation-dependent subunit trafficking. J Neurosci 26(17):4690-4700.

[86] Hanley MR. (1988) Proto-oncogenes in the nervous system. Neuron 1(3):175-182.

[87] Harder KW, Moller NP, Peacock JW, Jirik FR. (1998) Protein-tyrosine phosphatase alpha regulates Src family kinases and alters cell-substratum adhesion. J Biol Chem 273(48):31890-31900.

[88] Hartley DM, Kurth MC, Bjerkness L, Weiss JH, Choi DW. (1993) Glutamate receptor-induced Ca^{2+} accumulation in cortical cell culture correlates with subsequent neuronal degeneration. J Neurosci 13(5):1993-2000.

[89] Hasegawa S, Morioka M, Goto S, Korematsu K, Okamura A, Yano S, Kai Y, Hamada JI, Ushio Y. (2000) Expression of neuron specific phosphatase, striatal enriched phosphatase (STEP) in reactive astrocytes after transient forebrain ischemia. Glia 29(4):316-329.

[90] Hashimoto R, Fujimaki K, Jeong MR, Christ L, Chuang DM. (2003) Lithium-induced inhibition of Src tyrosine kinase in rat cerebral cortical neurons: a role in neuroprotection against N-methyl-D-aspartate receptor-mediated excitotoxicity. FEBS Lett 538(1-3):145-148.

[91] Hashimoto R, Hough C, Nakazawa T, Yamamoto T, Chuang DM. (2002) Lithium protection against glutamate excitotoxicity in rat cerebral cortical neurons: involvement of NMDA receptor inhibition possibly by decreasing NR2B tyrosine phosphorylation. J Neurochem 80(4):589-597.

[92] He J, Ogden LG, Vupputuri S, Bazzano LA, Loria C, Whelton PK. (1999) Dietary sodium intake and subsequent risk of cardiovascular disease in overweight adults. JAMA 282(21):2027-2034.

[93] He K, Rimm EB, Merchant A, Rosner BA, Stampfer MJ, Willett WC, Ascherio A. (2002) Fish consumption and risk of stroke in men. JAMA 288(24):3130-3136.

[94] Hisatsune C, Umemori H, Mishina M, Yamamoto T. (1999) Phosphorylation-dependent interaction of the N-methyl-D-aspartate receptor epsilon 2 subunit with phosphatidylinositol 3-kinase. Genes Cells 4(11):657-666.

[95] Horner RD, Matchar DB, Divine GW, Feussner JR. (1991) Racial variations in ischemic stroke-related physical and functional impairments. Stroke 22:1497-1501.

[96] Hou XY, Liu Y, Zhang GY. (2007) PP2, a potent inhibitor of Src family kinases, protects against hippocampal CA1 pyramidal cell death after transient global brain ischemia. Neurosci Lett 420(3):235-239.

[97] Hou XY, Zhang GY, Yan JZ, Chen M, Liu Y. (2002) Activation of NMDA receptors and L-type voltage-gated calcium channels mediates enhanced formation of Fyn-PSD95-NR2A complex after transient brain ischemia. Brain Res 955(1-2):123-132.

[98] Hou XY, Zhang GY, Zong YY. (2003) Suppression of postsynaptic density protein 95 expression attenuates increased tyrosine phosphorylation of NR2A subunits of N-methyl-D-aspartate receptors and interactions of Src and Fyn with NR2A after transient brain ischemia in rat hippocampus. Neurosci Lett 343(2):125-128.

[99] Hu BR, Wieloch T. (1994) Tyrosine phosphorylation and activation of mitogen-activated protein kinase in the rat brain following transient cerebral ischemia. J Neurochem 62(4):1357-1367.

[100] Hu NW, Ondrejcak T, Rowan MJ. (2012) Glutamate receptors in preclinical research on Alzheimer's disease: update on recent advances. Pharmacol Biochem Behav 100(4):855-862.

[101] Huang Y, Lu W, Ali DW, Pelkey KA, Pitcher GM, Lu YM, Aoto H, Roder JC, Sasaki T, Salter MW, MacDonald JF. (2001) CAKbeta/Pyk2 kinase is a signaling link for induction of long-term potentiation in CA1 hippocampus. Neuron 29(2):485-96.

[102] Ingraham CA, Cox ME, Ward DC, Fults DW, Maness PF. (1989) c-src and other proto-oncogenes implicated in neuronal differentiation. Mol Chem Neuropathol 10(1):1-14.

[103] Jiang X, Mu D, Biran V, Faustino J, Chang S, Rincón CM, Sheldon RA, Ferriero DM. (2008) Activated Src kinases interact with the N-methyl-D-aspartate receptor after neonatal brain ischemia. Ann Neurol 63(5):632-641.

[104] Jourdain P, Bergersen LH, Bhaukaurally K, Bezzi P, Santello M, Domercq M, Matute C, Tonello F, Gundersen V, Volterra A. (2007) Glutamate exocytosis from astrocytes controls synaptic strength. Nat Neurosci. 10(3):331-9.

[105] Kalia LV, Salter MW. (2003) Interactions between Src family protein tyrosine kinases and PSD-95. Neuropharmacology. 45(6):720-728.

[106] Kennedy MB. (1997) The postsynaptic density at glutamatergic synapses. Trends Neurosci 20(6):264-268.

[107] Kindy MS. (1993) Inhibition of tyrosine phosphorylation prevents delayed neuronal death following cerebral ischemia. J Cereb Blood Flow Metab 13(3):372-377.

[108] Kohr G, Seeburg P. (1996) Subtype-specific regulation of recombinant NMDA receptor channels by protein tyrosine kinases of the src family. J Physiol 492:445-452.

[109] Koistinaho J, Hökfelt T. (1997) Altered gene expression in brain ischemia. Neuroreport 8(2):i-viii.

[110] Komuro H, Rakic P. (1993) Modulation of neuronal migration by NMDA receptors. Science 260(5104):95-97.

[111] Kontos HA. (2001) Oxygen radicals in cerebral ischemia. The 2001 Willis Lecture. Stroke 32(11):2712-2715.

[112] Koritzinsky M, Wouters BG. (2007) Hypoxia and regulation of messenger RNA translation. Methods Enzymol 435:247-273.

[113] Kornau HC, Schenker LT, Kennedy MB, Seeburg PH. (1995) Domain interaction between NMDA receptor subunits and the postsynaptic density protein PSD-95. Science 269(5231):1737-1740.

[114] Krupp JJ, Vissel B, Thomas CG, Heinemann SF, Westbrook GL. (2002) Calcineurin acts via the C-terminus of NR2A to modulate desensitization of NMDARs. Neuropharmacology 42:593-602.

[115] Kudo Y, Ogura A. (1986) Glutamate-induced increase in intracellular Ca²⁺ concentration in isolated hippocampal neurones. Br J Pharmacol 89(1):191-198.

[116] Kuo WL, Chung KC, Rosner MR. (1997) Differentiation of central nervous system neuronal cells by fibroblast-derived growth factor requires at least two signaling pathways: roles for Ras and Src. Mol Cell Biol 17(8):4633-4643.

[117] Lafon-Cazal M, Pietri S, Culcasi M, Bockaert J. (1993) NMDA-dependent superoxide production and neurotoxicity. Nature 364(6437):535-537.

[118] Lan JY, Skeberdis VA, Jover T, Grooms SY, Lin Y, Araneda RC, Zheng X, Bennett MV, Zukin RS. (2001) Protein kinase C modulates NMDA receptor trafficking and gating. Nat Neurosci. 4(4):382-390.

[119] Lau A, Tymianski M. (2010) Glutamate receptors, neurotoxicity and neurodegeneration. Pflugers Arch 460(2):525-542.

[120] Lau C, Zukin R. (2007) NMDA receptor trafficking in synaptic plasticity and neuropsychiatric disorders. Nat Rev Neurosci 8(6):413-426.

[121] Lau L, Huganir R. (1995) Differential tyrosine phosphorylation of N-methyl-D-aspartate receptor subunits. J Biol Chem 270(34):20036-20041.

[122] Lauri SE, Taira T, Rauvala H. (2000) High-frequency synaptic stimulation induces association of fyn and c- src to distinct phosphorylated components. NeuroReport 11:997-1000.

[123] Lavezzari G, McCallum J, Lee R, Roche KW. (2003) Differential binding of the AP-2 adaptor complex and PSD-95 to the C-terminus of the NMDA receptor subunit NR2B regulates surface expression. Neuropharmacology 45(6):729-737.

[124] Lavezzari G, McCallum J, Dewey CM, Roche KW. (2004) Subunit-specific regulation of NMDA receptor endocytosis. J Neurosci 24(28):6383-6391.

[125] Lee HK. (2006) Synaptic plasticity and phosphorylation. Pharmacol Therapeutics 112:810-832.

[126] Lees GJ. (1991) Inhibition of sodium-potassium-ATPase: a potentially ubiquitous mechanism contributing to central nervous system pathology. Brain Res Rev 16(3):283-300.

[127] Lei G, Xue S, Chéry N, Liu Q, Xu J, Kwan CL, Fu YP, Lu YM, Liu M, Harder KW, Yu XM. (2002) Gain control of N-methyl-D-aspartate receptor activity by receptor-like protein tyrosine phosphatase alpha. EMBO J 21(12):2977-2989.

[128] Le Meur K, Galante M, Angulo MC, Audinat E. (2007) Tonic activation of NMDA receptors by ambient glutamate of non-synaptic origin in the rat hippocampus. J Physiol. 580(Pt. 2):373-83.

[129] Lev S, Moreno H, Martinez R, Canoll P, Peles E, Musacchio JM, Plowman GD, Rudy B, Schlessinger J. (1995) Protein tyrosine kinase PYK2 involved in Ca(2+)-induced regulation of ion channel and MAP kinase functions. Nature 376(6543):737-745.

[130] Li T, Yu XJ, Zhang GY. (2008) Tyrosine phosphorylation of HPK1 by activated Src promotes ischemic brain injury in rat hippocampal CA1 region. FEBS Lett 582(13):1894-1900.

[131] Li Y, Mu Y, Gage FH. (2009) Development of neural circuits in the adult hippocampus. Curr Top Dev Biol 87:149-174.

[132] Lin Y, Skeberdis VA, Francesconi A, Bennett MV, Zukin RS. (2004) Postsynaptic density protein-95 regulates NMDA channel gating and surface expression. J Neurosci 24(45):10138-10148.

[133] Lipton P (1999) Ischemic cell death in brain neurons. Physiol Rev 79(4):1431-1568.

[134] Liu X, Brodeur SR, Gish G, Songyang Z, Cantley LC, Laudano AP, Pawson T. (1993) Regulation of c-Src tyrosine kinase activity by the Src SH2 domain. Oncogene 8(5):1119-1126.

[135] Liu Y, Zhang G, Gao C, Hou X. (2001) NMDA receptor activation results in tyrosine phosphorylation of NMDA receptor subunit 2A(NR2A) and interaction of Pyk2 and Src with NR2A after transient cerebral ischemia and reperfusion. Brain Res 909(1-2):51-58.

[136] Liu Y, Zhang GY, Yan JZ, Xu TL. (2005) Suppression of Pyk2 attenuated the increased tyrosine phosphorylation of NMDA receptor subunit 2A after brain ischemia in rat hippocampus. Neurosci Lett 379(1):55-58.

[137] Lobner D, Lipton P. (1993) Intracellular calcium levels and calcium fluxes in the CA1 region of the rat hippocampal slice during in vitro ischemia: relationship to electrophysiological cell damage. J Neurosci 13:4861-4871.

[138] Lombroso PJ, Murdoch G, Lerner M. (1991) Molecular characterization of a protein-tyrosine-phosphatase enriched in striatum. Proc Natl Acad Sci USA 88(16):7242-7246.

[139] Love S. (1999) Oxidative stress in brain ischemia. Brain Pathol 9(1):119-131.

[140] Lu WY, Xiong ZG, Lei S, Orser BA, Dudek E, Browning MD, MacDonald JF. (1999) G-protein-coupled receptors act via protein kinase C and Src to regulate NMDA receptors. Nat Neurosci 2(4):331-338.

[141] Lu YM, Roder JC, Davidow J, Salter MW. (1998) Src activation in the induction of long-term potentiation in CA1 hippocampal neurons. Science 279(5355):1363-1367.

[142] Lucas DR, Newhouse JP. (1957) The toxic effects of sodium L-glutamate on the inner layer of the retina. Arch Ophthalm 58:193-201.

[143] Luiten PG, Stuiver B, de Jong GI, Nyakas C, De Keyser JH. (1997) Calcium homeostasis, nimodipine, and stroke. In Clinical Pharmacology of Cerebral Ischemia, Humana Press, New Jersey, pp. 67-99.

[144] Ma J, Zhang GY. (2003) Lithium reduced N-methyl-D-aspartate receptor subunit 2A tyrosine phosphorylation and its interactions with Src and Fyn mediated by PSD-95 in rat hippocampus following cerebral ischemia. Neurosci Lett 348(3):185-189.

[145] Ma J, Zhang GY, Liu Y, Yan JZ, Hao ZB. (2004) Lithium suppressed Tyr-402 phosphorylation of proline-rich tyrosine kinase (Pyk2) and interactions of Pyk2 and PSD-95 with NR2A in rat hippocampus following cerebral ischemia. Neurosci Res 49(4):357-362.

[146] MacDonald JF, Jackson MF, Beazely MA. (2006) Hippocampal long-term synaptic plasticity and signal amplification of NMDA receptors. Crit Rev Neurobiol 18(1-2):71-84.

[147] MacDonald JF, Kotecha SA, Lu WY, Jackson MF. (2001) Convergence of PKC-dependent kinase signal cascades on NMDA receptors. Curr Drug Targets 2(3):299-312.

[148] MacDonald JF, Mody I, Salter MW. (1989) Regulation of N-methyl-D-aspartate receptors revealed by intracellular dialysis of murine neurones in culture. J Physiol 414:17-34.

[149] Machado-Vieira R, Manji HK, Zarate CA. (2009) The role of the tripartite glutamatergic synapse in the pathophysiology and therapeutics of mood disorders. Neuroscientist 15(5):525-39.

[150] Machida K, Mayer BJ. (2005) The SH2 domain: versatile signaling module and pharmaceutical target. Biochim Biophys Acta 1747(1):1-25.

[151] Martone ME, Hu BR, Ellisman MH. (2000) Alterations of hippocampal postsynaptic densities following transient ischemia. Hippocampus 10(5):610-616.

[152] Mayer ML, Westbrook GL. (1987) Permeation and block of N-methyl-D-aspartic acid receptor channels by divalent cations in mouse cultured central neurones. J Physiol 394:501-527.

[153] Mayer M, Armstrong N. (2004) Structure and function of glutamate receptor ion channels. Annu Rev Physiol 66:161-181.

[154] McLennan H. (1983) Receptors for the excitatory amino acids in the mammalian central nervous system. Prog Neurobiol 20(3-4):251-271.

[155] Meldrum BS. (1990) Protection against ischemic neuronal damage by drugs acting on excitatory neurotransmission. Cerebrovasc Brain Metab Rev 2:27-57.

[156] Monyer H, Burnashev N, Laurie DJ, Sakmann B, Seeburg PH. (1994) Developmental and regional expression in the rat brain and functional properties of four NMDA receptors. Neuron 12:529–540.

[157] Moon IS, Apperson ML, Kennedy MB. (1994) The major tyrosine-phosphorylated protein in the postsynaptic density fraction is N-methyl-D-aspartate receptor subunit 2B. Proc Natl Acad Sci USA 91(9):3954-3958.

[158] Moriyoshi K, Masu M, Ishii T, Shigemotto R, Mizuno N, Nakanishi S. (1991) Molecular cloning and characterization of the rat NMDA receptor. Nature 354:31-37.

[159] Moss SJ, Gorrie GH, Amato A, Smart TG. (1995) Modulation of GABAA receptors by tyrosine phosphorylation. Nature 377(6547):344-348.

[160] Murotomi K, Takagi N, Takayanagi G, Ono M, Takeo S, Tanonaka K. (2008) mGluR1 antagonist decreases tyrosine phosphorylation of NMDA receptor and attenuates infarct size after transient focal cerebral ischemia. J Neurochem 105(5):1625-1634.

[161] Nada S, Shima T, Yanai H, Husi H, Grant SG, Okada M, Akiyama T. (2003) Identification of PSD-93 as a substrate for the Src family tyrosine kinase Fyn. J Biol Chem 278(48):47610-47621.

[162] Naegele JR, Lombroso PJ. (1994) Protein tyrosine phosphatases in the nervous system. Crit Rev Neurobiol 9(1):105-114.

[163] Nakazawa, T, Komai S, Tezuka T, Hisatsune C, Umemori H, Semba K, Mishina M, Manabe T, Yamamoto T. (2001) Characterization of Fyn-mediated tyrosine phosphorylation sites on GluR 2 (NR2B) subunit of the N-methyl-D-aspartate receptor. J Biol Chem 276: 693-699.

[164] Nakazawa T, Komai S, Watabe AM, Kiyama Y, Fukaya M, Arima-Yoshida F, Horai R, Sudo K, Ebine K, Delaware M, Goto J, Umemori H, Tezuka T, Iwakura Y, Watanabe M,

Yamamoto T, Manabe T. (2006) NR2B tyrosine phosphorylation modulates fear learning as well as amygdaloid synaptic plasticity. EMBO J 25(12):2867-2877.

[165] Nguyen TH, Liu J, Lombroso PJ. (2002) Striatal enriched phosphatase 61 dephosphorylates Fyn at phosphotyrosine 420. J Biol Chem 277(27):24274-24279.

[166] Nguyen TH, Paul S, Xu Y, Gurd JW, Lombroso PJ. (1999) Calcium-dependent cleavage of striatal enriched tyrosine phosphatase (STEP). J Neurochem 73(5):1995-2001.

[167] Niethammer M, Kim E, Sheng M. (1996) Interaction between the C terminus of NMDA receptor subunits and multiple members of the PSD-95 family of membrane-associated guanylate kinases. J Neurosci 16(7):2157-2163.

[168] Nowak L, Bregestovski P, Ascher P, Herbet A, Prochiantz A. (1984) Magnesium gates glutamate-activated channels in mouse central neurones. Nature 307(5950):462-465.

[169] O'Collins VE, Macleod MR, Donnan GA, Horky LL, van der Worp BH, Howells DW. (2006) 1,026 experimental treatments in acute stroke. Ann Neurol 59(3):467-477.

[170] Ohnishi H, Murata Y, Okazawa H, Matozaki T. (2011) Src family kinases: modulators of neurotransmitter function and behaviour. Trends Neurosci 34(12):629-637.

[171] Ohtsuki T, Matsumoto M, Kitagawa K, Mabuchi T, Mandai K, Matsushita K, Kuwabara K, Tagaya M, Ogawa S, Ueda H, Kamada T, Yanagihara T. (1996) Delayed neuronal death in ischemic hippocampus involves stimulation of protein tyrosine phosphorylation. Am J Physiol 271(4 Pt 1):C1085-1097.

[172] Olesen J, Gustavsson A, Svensson M, Wittchen HU, Jönsson B. (2012) The economic cost of brain disorders in Europe. Eur J Neurol 19(1):155-162.

[173] Olney JW. (1969) Glutamate-induced retinal degeneration in neonatal mice. Electron microscopy of the acutely evolving lesion. J Neuropathol Exp Neurol 28(3):455-474.

[174] Oyama T, Goto S, Nishi T, Sato K, Yamada K, Yoshikawa M, Ushio Y. (1995) Immunocytochemical localization of the striatal enriched protein tyrosine phosphatase in the rat striatum: a light and electron microscopic study with a complementary DNA-generated polyclonal antibody. Neuroscience 69(3):869-880.

[175] Ozawa S, Kamiya H, Tsuzuki K. (1998) Glutamate receptors in the mammalian central nervous system. Prog Neurobiol 54:581-618.

[176] Pallen CJ. (2003) Protein tyrosine phosphatase alpha (PTPalpha): a Src family kinase activator and mediator of multiple biological effects. Curr Top Med Chem 3(7):821-835.

[177] Paoletti P, Ascher P, Neyton J. (1997) High-affinity zinc inhibition of NMDA NR1-NR2A receptors. J Neurosci 17(15):5711-25.

[178] Pawson T, Scott JD. (2005) Protein phosphorylation in signaling - 50 years and counting. Trends Biochem Sci (6):286-290.

[179] Pei L, Li Y, Zhang GY, Cui ZC, Zhu ZM. (2000) Mechanisms of regulation of tyrosine phosphorylation of NMDA receptor subunit 2B after cerebral ischemia/reperfusion. Acta Pharmacol Sin 21(8):695-700.

[180] Pei L, Li Y, Yan JZ, Zhang GY, Cui ZC, Zhu ZM. (2000) Changes and mechanisms of protein-tyrosine kinase and protein-tyrosine phosphatase activities after brain ischemia/reperfusion. Acta Pharmacol Sin 21(8):715-720.

[181] Pelkey K, Askalan R, Paul S, Kalia L, Nguyen T, Pitcher G, Salter M, Lombroso P. (2002) Tyrosine phosphatase STEP is a tonic brake on induction of long-term potentiation. Neuron 34: 127-138.

[182] Perkinton MS, Ip JK, Wood GL, Crossthwaite AJ, Williams RJ. (2002) Phosphatidylinositol 3-kinase is a central mediator of NMDA receptor signalling to MAP kinase (Erk1/2), Akt/PKB and CREB in striatal neurones. J Neurochem 80(2):239-254.

[183] Petralia RS, Al-Hallaq R, Wenthold RJ. (2009) Trafficking and Targeting of NMDA receptors. In Biology of the NMDA Receptor, CRC Press, New York, pp. 149-200.

[184] Petralia RS, Sans N, Wang YX, Wenthold RJ. (2005) Ontogeny of postsynaptic density proteins at glutamatergic synapses. Mol Cell Neurosci 29:436-452.

[185] Phillis JW, O'Regan MH. (2004) A potentially critical role of phospholipases in central nervous system ischemic, traumatic, and neurodegenerative disorders. Brain Res Brain Res Rev 44(1):13-47.

[186] Ponniah S, Wang DZ, Lim KL, Pallen CJ. (1999) Targeted disruption of the tyrosine phosphatase PTPalpha leads to constitutive downregulation of the kinases Src and Fyn. Curr Biol 9(10):535-538.

[187] Powers WJ. (1990) Stroke. In Neurobiology of Disease, Oxford University Press, Toronto, pp. 339-355.

[188] Prybylowski K, Chang K, Sans N, Kan L, Vicini S, Wenthold RJ. (2005) The synaptic localization of NR2B-containing NMDA receptors is controlled by interactions with PDZ proteins and AP-2. Neuron 47(6):845-857.

[189] Qizilbash N, Duffy SW, Warlow C, Mann, J. (1992) Lipids are risk factors for ischemic stroke – review and overview. Cerebrovascular Disease 2:127-136.

[190] Randall RD, Thayer SA. (1992) Glutamate-induced calcium transient triggers delayed calcium overload and neurotoxicity in rat hippocampal neurons. J Neurosci 12:1882-1895.

[191] Rao A, Craig AM. (1997) Activity regulates the synaptic localization of the NMDA receptor in hippocampal neurons. Neuron 19(4):801-812.

[192] Ray SK. (2006) Currently evaluated calpain and caspase inhibitors for neuroprotection in experimental brain ischemia. Curr Med Chem 13(28):3425-3440.

[193] Récasens M, Guiramand J, Aimar R, Abdulkarim A, Barbanel G. (2007) Metabotropic glutamate receptors as drug targets. Curr Drug Targets 8(5):651-681.

[194] Reynolds IJ, Hastings TG. (1995) Glutamate induces the production of reactive oxygen species in cultured forebrain neurons following NMDA receptor activation. J Neurosci 15:3318-3327.

[195] Roche S, Koegl M, Barone MV, Roussel MF, Courtneidge SA. (1995) DNA synthesis induced by some but not all growth factors requires Src family protein tyrosine kinases. Mol Cell Biol 15(2):1102-1109.

[196] Roche KW, Standley S, McCallum J, Dune Ly C, Ehlers MD, Wenthold RJ. (2001) Molecular determinants of NMDA receptor internalization. Nat Neurosci 4(8):794-802.

[197] Rong Y, Lu X, Bernard A, Khrestchatisky M, Baudry M. (2001) Tyrosine phosphorylation of ionotropic glutamate receptors by Fyn or Src differentially

modulates their susceptibility to calpain and enhances their binding to spectrin and PSD-95. J Neurochem 79(2):382-390.

[198] Rosamond W, Flegal K, Friday G, et al. (2007) Heart disease and stroke statistics – 2007 update: a report from the American heart association statistics committee and stroke statistics subcommittee. Circulation 115, e69–e110.

[199] Rothman SM Olney JW. (1986) Glutamate and the pathophysiology of hypoxic-ischemic brain damage. Ann Neurol 19:105-111.

[200] Rowe MK, Chuang DM. (2004) Lithium neuroprotection: molecular mechanisms and clinical implications. Expert Rev Mol Med 6(21):1-18.

[201] Ryan TJ, Emes RD, Grant SG, Komiyama NH. (2008) Evolution of NMDA receptor cytoplasmic interaction domains: implications for organisation of synaptic signalling complexes. BMC Neurosci 9:6.

[202] Sacco RL, Boden-Albala B, Gan R, Chen X, Kargman DE, Shea S, Paik MC, Hauser WA. (1998) Stroke incidence among white, black, and Hispanic residents of an urban community. The Northern Manhattan Stroke Study. Am J Epidemiology 147:259-268.

[203] Sacco RL, Boden-Albala B. (2001) Stroke risk factors: identifiction and modification. In Stroke Therapy, 2nd Edition, Butterworth-Heinemann, U.S.A., pp. 1-24.

[204] Salter M, Kalia L. (2004) Src kinases: a hub for NMDA receptor regulation. Nat Rev Neurosci 5:317-328.

[205] Salter MW, Dong Y, Kalia LV, Liu XJ, Pitcher G. (2009) Regulation of NMDA receptors by kinases and phosphatases. In Biology of the NMDA Receptor, CRC Press, New York, pp. 123-148.

[206] Sato Y, Tao YX, Su Q, Johns RA. (2008) Post-synaptic density-93 mediates tyrosine phosphorylation of the N-methyl-D-aspartate receptors. Neuroscience 153(3):700-708.

[207] Sattler R, Charlton MP, Hafner M, Tymianski M. (1998) Distinct influx pathways, not calcium load, determine neuronal vulnerability to calcium neurotoxicity. J Neurochem 71(6):2349-2364.

[208] Schlaepfer WW, Bunge RP. (1973) Effects of calcium ion concentration on the degeneration of amputated axons in tissue culture. J Cell Biol 59:456-470.

[209] Shamloo M, Wieloch T. (1999) Changes in protein tyrosine phosphorylation in the rat brain after cerebral ischemia in a model of ischemic tolerance. J Cereb Blood Flow Metab 19(2):173-183.

[210] Shaper AG, Phillips AN, Pocock SJ, Walker M, Macfarlane PW. (1991) Risk factors for stroke in middle aged British men. BMJ 302(6785):1111-5.

[211] Sicheri F, Kuriyan J. (1997) Structures of Src-family tyrosine kinases. Curr Opin Struct Biol 7(6):777-785.

[212] Siciliano JC, Toutant M, Derkinderen P, Sasaki T, Girault JA. (1996) Differential regulation of proline-rich tyrosine kinase 2/cell adhesion kinase beta (PYK2/CAKbeta) and pp125(FAK) by glutamate and depolarization in rat hippocampus. J Biol Chem 271(46):28942-28946.

[213] Siesjo BK. (1986) Calcium and ischemic brain damage. Exp Neurol 25(Suppl 1):45-56.

[214] Siesjo BK, Katsura K, Mellergard P, Ekholm A, Lundgren J, Smith ML. (1993) Acidosis-related brain damage. Prog Brain Res 96:23-48.

[215] Simon RP, Swan JH, Griffiths T, Meldrum BS. (1984) Blockade of NMDA receptors may protect against ischemic damage in the brain. Science 226:850-852.

[216] Snyder EM, Nong Y, Almeida CG, Paul S, Moran T, Choi EY, Nairn AC, Salter MW, Lombroso PJ, Gouras GK, Greengard P. (2005) Regulation of NMDA receptor trafficking by amyloid-beta. Nat Neurosci 8(8):1051-1058.

[217] Sorimachi H, Ishiura S, Suzuki K. (1997) Structure and physiological function of calpains. Biochem J 328:721-732.

[218] Stehelin D, Varmus HE, Bishop JM, Vogt PK. (1976) DNA related to the transforming gene(s) of avian sarcoma viruses is present in normal avian DNA. Nature 260(5547):170-173.

[219] Stoker A, Dutta R. (1998) Protein tyrosine phosphatases and neural development. Bioessays 20(6):463-472.

[220] Sudol M, Grant SG, Maisonpierre PC. (1993) Proto-oncogenes and signaling processes in neural tissues. Neurochem Int 4:369-384.

[221] Suzuki T, Okumura-Noji K. (1995) NMDA receptor subunits epsilon 1 (NR2A) and epsilon 2 (NR2B) are substrates for Fyn in the postsynaptic density fraction isolated from the rat brain. Biochem Biophys Res Commun 216(2):582-588.

[222] Swope SL, Moss SJ, Raymond LA, Huganir RL. (1999) Regulation of ligand-gated ion channels by protein phosphorylation. Adv Second Messenger Phosphoprotein Res 33:49-78.

[223] Szydlowska K, Tymianski M. (2010) Calcium, ischemia and excitotoxicity. Cell Calcium 47(2):122-129.

[224] Takagi K, Ginsberg MD, Globus MY, Dietrich WD, Martinez E, Kraydieh S, Busto R. (1993) Changes in amino acid neurotransmitters and cerebral blood flow in the ischemic penumbral region following middle cerebral artery occlusion in the rat: correlation with histopathology. J Cereb Blood Flow Metab 13(4):575-585.

[225] Takagi N, Shinno K, Teves L, Bissoon N, Wallace MC, Gurd JW. (1997) Transient ischemia differentially increases tyrosine phosphorylation of NMDA receptor subunits 2A and 2B. J Neurochem 69(3):1060-1065.

[226] Takagi N, Cheung HH, Bissoon N, Teves L, Wallace MC, Gurd JW. (1999) The effect of transient global ischemia on the interaction of Src and Fyn with the N-methyl-D-aspartate receptor and postsynaptic densities: possible involvement of Src homology 2 domains. J Cereb Blood Flow Metab 19(8):880-888.

[227] Takagi N, Logan R, Teves L, Wallace MC, Gurd JW. (2000) Altered interaction between PSD-95 and the NMDA receptor following transient global ischemia. J Neurochem 74(1):169-178.

[228] Takagi N, Sasakawa K, Besshoh S, Miyake-Takagi K, Takeo S. (2003) Transient ischemia enhances tyrosine phosphorylation and binding of the NMDA receptor to the Src homology 2 domain of phosphatidylinositol 3-kinase in the rat hippocampus. J Neurochem 84(1):67-76.

[229] Tanaka C, Nishizuka Y. (1994) The protein kinase C family for neuronal signaling. Annu Rev Neurosci 17:551-567.

[230] Ter Horst G, Postigo A. (1997) Stroke: prevalence and mechanisms of cell death. In Clinical Pharmacology of Cerebral Ischaemia, Humana, New Jersey, pp. 1-30.

[231] Tezuka T, Umemori H, Akiyama T, Nakanishi S, Yamamoto T. (1999) PSD-95 promotes Fyn-mediated tyrosine phosphorylation of the N-methyl-D-aspartate receptor subunit NR2A. Proc Natl Acad Sci USA 96(2):435-440.

[232] Thayer SA, Usachev YM, Pottorf WJ. (2002) Modulating Ca^{2+} clearance from neurons. Frontiers Biosci 7:d1255-1279.

[233] Tovar KR, Westbrook GL (1999) The incorporation of NMDA receptors with a distinct subunit composition at nascent hippocampal synapses in vitro. J Neurosci 19:4180-4188.

[234] Tovar KR, Westbrook GL. (2002) Mobile NMDA receptors at hippocampal synapses. Neuron 34(2):255-264.

[235] Traub LM. (2009) Tickets to ride: selecting cargo for clathrin-regulated internalization. Nat Rev Mol Cell Biol 10(9):583-596.

[236] Traynelis SF, Wollmuth LP, McBain CJ, Menniti FS, Vance KM, Ogden KK, Hansen KB, Yuan H, Myers SJ, Dingledine R. (2010) Glutamate receptor ion channels: structure, regulation, and function. Pharmacol Rev 62(3):405-496.

[237] Trepanier CH, Jackson MF, MacDonald JF. (2012) Regulation of NMDA receptors by the tyrosine kinase Fyn. FEBS Journal 279:12-19.

[238] Ubersax JA, Ferrell JE Jr. (2007) Mechanisms of specificity in protein phosphorylation. Nat Rev Mol Cell Biol 8(7):530-541.

[239] Umemori H, Ogura H, Tozawa N, Mikoshiba K, Nishizumi H, Yamamoto T. (2003) Impairment of N-methyl-D-aspartate receptor-controlled motor activity in LYN-deficient mice. Neuroscience 118(3):709-713.

[240] Vaughan J & Bullock R. (1999) Cellular and vascular pathophysiology of stroke. In Stroke Therapy: Basic, Preclinical, and Clinical Directions, Wiley-Liss, Toronto, pp. 3-38.

[241] Vinade L, Petersen JD, Do K, Dosemeci A, Reese TS. (2001) Activation of calpain may alter the postsynaptic density structure and modulate anchoring of NMDA receptors. Synapse 40(4):302-309.

[242] Vissel B, Krupp JJ, Heinemann SF, Westbrook GL. (2001) A use-dependent tyrosine dephosphorylation of NMDA receptors is independent of ion flux. Nat Neurosci 4(6):587-596.

[243] Waksman G, Kumaran S, Lubman O. (2004) SH2 domains: role, structure and implications for molecular medicine. Expert Rev Mol Med 6(3):1-18.

[244] Walaas SI, Greengard P. (1991) Protein phosphorylation and neuronal function. Pharmacol Rev 43(3):299-349.

[245] Wan Q, Xiong ZG, Man HY, Ackerley CA, Braunton J, Lu WY, Becker LE, MacDonald JF, Wang YT. (1997) Recruitment of functional GABA(A) receptors to postsynaptic domains by insulin. Nature 388(6643):686-690.

[246] Wang K, Hackett JT, Cox ME, Van Hoek M, Lindstrom JM, Parsons SJ. (2004) Regulation of the neuronal nicotinic acetylcholine receptor by SRC family tyrosine kinases. J Biol Chem. 279(10):8779-8786.

[247] Wang WQ, Sun JP, Zhang ZY. (2003) An overview of the protein tyrosine phosphatase superfamily. Curr Top Med Chem 3(7):739-748.

[248] Wang WW, Hu SQ, Li C, Zhou C, Qi SH, Zhang GY. (2010) Transduced PDZ1 domain of PSD-95 decreases Src phosphorylation and increases nNOS (Ser847) phosphorylation contributing to neuroprotection after cerebral ischemia. Brain Res 1328:162-170.

[249] Wang Y, Salter M. (1994) Regulation of NMDA receptors by tyrosine kinases and phosphatases. Nature 369:233-235.

[250] Wang Y, Yu X, Salter M. (1996) Ca^{2+}-independent reduction of N-methyl-D-aspartate channel activity by protein tyrosine phosphatase. Proc Natl Acad Sci USA 93:1721-1725.

[251] Wannamethee G, Shaper AG. (1992) Physical activity and stroke in British middle aged men. BMJ 304:597-601.

[252] Watanabe M, Inoue Y, Sakimura K, Mishina M. (1992) Developmental changes in distribution of NMDA receptor channel subunit mRNAs. NeuroReport 3:1138-1140.

[253] Watanabe M, Inoue Y, Sakimura K, Mishina M. (1993) Distinct distributions of five N-methyl-D-aspartate receptor channel subunit mRNAs in the forebrain. J Comp Neurol 338(3):377-390.

[254] Wei G, Yin Y, Li W, Bito H, She H, Mao Z. (2012) Calpain-mediated degradation of myocyte enhancer factor 2D contributes to excitotoxicity by activation of extrasynaptic N-methyl-D-aspartate receptors. J Biol Chem 287(8):5797-5805.

[255] World Health Organization. Global Burden of Stroke. (2004) The atlas of heart disease and   stroke. http://www.who.int/cardiovascular_diseases/en/cvd_atlas_15_burden_stroke.pdf.

[256] Wolf PA, D'Agostino RB, Belanger AJ, Kannel WB. (1991a) Probability of stroke: a risk profile from the Framingham study. Stroke 22:312-318.

[257] Wolf PA, Abbott RD, Kannel WB. (1991b) Atrial fibrillation as an independent risk factor for stroke: the Framingham study. Stroke 22:983-988.

[258] Wu HW, Li HF, Guo J. (2007) N-methyl-D-aspartate receptors mediate diphosphorylation of extracellular signal-regulated kinases through Src family tyrosine kinases and Ca2+/calmodulin-dependent protein kinase II in rat hippocampus after cerebral ischemia. Neurosci Bull 23(2):107-112.

[259] Xiong WC, Mei L. (2003) Roles of FAK family kinases in nervous system. Front Biosci 8:s676-s682.

[260] Xiong Z, Raouf R, Lu WY, Wang LY, Orser BA, Dudek EM, Browning MD, MacDonald JF. (1998) Regulation of N-methyl-D-aspartate receptor function by constitutively active protein kinase C. Mol Pharmacol 54(6):1055-1063.

[261] Xiong Z, Pelkey K, Lu W, Lu Y, Roder J, MacDonald J, Salter M. (1999) Src potentiation of NMDA receptors in hippocampal and spinal neurons is not mediated by reducing zinc inhibition. J Neurosci 19:RC37(1-6).

[262] Xu J, Kurup P, Bartos JA, Patriarchi T, Hell JW, Lombroso PJ. (2012) Striatal-enriched Protein-tyrosine Phosphatase (STEP) Regulates Pyk2 Kinase Activity. J Biol Chem 287(25):20942-20956.

[263] Xu J, Kurup P, Zhang Y, Goebel-Goody SM, Wu PH, Hawasli AH, Baum ML, Bibb JA, Lombroso PJ. (2009) Extrasynaptic NMDA receptors couple preferentially to excitotoxicity via calpain-mediated cleavage of STEP. J Neurosci 29(29):9330-9343.

[264] Xu J, Liu Y, Zhang GY. (2008) Neuroprotection of GluR5-containing kainate receptor activation against ischemic brain injury through decreasing tyrosine phosphorylation of N-methyl-D aspartate receptors mediated by Src kinase. J Biol Chem 283(43):29355-29366.

[265] Xu W. (2011) PSD-95-like membrane associated guanylate kinases (PSD-MAGUKs) and synaptic plasticity. Curr Opin Neurobiol 21(2):306-312.

[266] Xu W, Doshi A, Lei M, Eck MJ, Harrison SC. (1999) Crystal structures of c-Src reveal features of its autoinhibitory mechanism. Mol Cell 3(5):629-638.

[267] Yaka R, Thornton C, Vagts AJ, Phamluong K, Bonci A, Ron D. (2002) NMDA receptor function is regulated by the inhibitory scaffolding protein, RACK1. Proc Natl Acad Sci USA 99(8):5710-57155.

[268] Yan JZ, Liu Y, Zong YY, Zhang GY. (2012) Knock-down of postsynaptic density protein 95 expression by antisense oligonucleotides protects against apoptosis-like cell death induced by oxygen-glucose deprivation in vitro. Neurosci Bull 28(1):69-76.

[269] Yokota M, Saido TC, Miyaji K, Tani E, Kawashima S, Suzuki K. (1994) Stimulation of protein-tyrosine phosphorylation in gerbil hippocampus after global forebrain ischemia. Neurosci Lett 168(1-2):69-72.

[270] Yu XM, Askalan R, Keil GJ, Salter MW. (1997) NMDA channel regulation by channel associated protein tyrosine kinase Src. Science 275(5300):674-678.

[271] Zalewska T, Ziemka-Nałecz M, Domańska-Janik K. (2005) Transient forebrain ischemia effects interaction of Src, FAK, and PYK2 with the NR2B subunit of N-methyl-D-aspartate receptor in gerbil hippocampus. Brain Res 1042(2):214-223.

[272] Zhang C, Siman R, Xu YA, Mills AM, Frederick JR, Neumar RW. (2002) Comparison of calpain and caspase activities in the adult rat brain after transient forebrain ischemia. Neurobiol Dis 10(3):289-305.

[273] Zhang F, Li C, Wang R, Han D, Zhang QG, Zhou C, Yu HM, Zhang GY. (2007) Activation of GABA receptors attenuates neuronal apoptosis through inhibiting the tyrosine phosphorylation of NR2A by Src after cerebral ischemia and reperfusion. Neuroscience 150(4):938-949.

[274] Zhang L, Hsu JC, Takagi N, Gurd JW, Wallace MC, Eubanks JH. (1997) Transient global ischemia alters NMDA receptor expression in rat hippocampus: correlation with decreased immunoreactive protein levels of the NR2A/2B subunits, and an altered NMDA receptor functionality. J Neurochem 69(5):1983-1994.

[275] Zhang J, Diamond JS. (2006) Distinct perisynaptic and synaptic localization of NMDA and AMPA receptors on ganglion cells in rat retina. J Comp Neurol 498:810-820.

[276] Zhang Y, Lipton P. (1999) Cytosolic Ca^{2+} changes during in vitro ischemia in rat hippocampal slices: major roles for glutamate and Na^+ dependent Ca^{2+} release from mitochondria. J Neurosci 19(9):3307-3315.

[277] Zhang Y, Kurup P, Xu J, Carty N, Fernandez SM, Nygaard HB, Pittenger C, Greengard P, Strittmatter SM, Nairn AC, Lombroso PJ. (2010) Genetic reduction of striatal-enriched tyrosine phosphatase (STEP) reverses cognitive and cellular deficits in an Alzheimer's disease mouse model. Proc Natl Acad Sci USA 107(44):19014-19019.

[278] Zhao W, Cavallaro S, Gusev P, Alkon DL. (2000) Nonreceptor tyrosine protein kinase pp60c-src in spatial learning: synapse-specific changes in its gene expression, tyrosine phosphorylation, and protein-protein interactions. Proc Natl Acad Sci USA 97(14):8098-8103.

[279] Zheng XM, Wang Y, Pallen CJ. (1992) Cell transformation and activation of pp60c-src by overexpression of a protein tyrosine phosphatase. Nature 359(6393):336-339.

[280] Ziemka-Nalecz M, Zalewska T, Zajac H, Domansk-Janik K. (2003) Decrease of PKC precedes other cellular signs of calpain activation in area CA1 of the hippocampus after transient cerebral ischemia. Neurochem Int 42(3):205-214.

Chapter 2

Role of Tyrosine Kinase A Receptor (TrkA) on Pathogenicity of *Clostridium perfringens* Alpha-Toxin

Masataka Oda, Masahiro Nagahama, Keiko Kobayashi and Jun Sakurai

Additional information is available at the end of the chapter

1. Introduction

Clostridium perfringens (*C. perfringens*) is a toxin-producing anaerobic Gram-positive bacterium, which is well known for its role in human tissue infections and food poisoning. It is readily isolated from soil and a component of normal human intestinal and vaginal flora in many individuals. Apart from the classic clostridial myonecrosis of gas gangrene, *C. perfringens* can be responsible for a range of other clinical scenarios including sepsis, aspiration pneumonia, brain abscess, and enteritis necroticans. The potent exotoxins produced by various strains of *C. perfringens* are central to their effectiveness as pathogens, and include four major toxins used in strain classification: a phospholipase C (alpha-toxin, PLC), two pore-forming toxins (beta and epsilon toxins); and an ADP-ribosylation toxin (iota toxin). *C. perfringens* gas gangrene is one of the most fulminant necrotizing infections affecting humans. The infection can become well established in traumatized tissues in as little as 6-8 h and the destruction of adjacent healthy muscle can progress several inches per hour despite appropriate antibiotic coverage. Shock and organ failure occur in 50% of patients, and 40% of these individuals die. Even with modern medical advances and intensive care regimens, the centuries-old practice of radical amputation on an emergent basis remains the single best treatment. Histologically, this infection is characterized by widespread destruction of muscle and the absence of polymorphonuclear leukocytes at the site of infection. Instead, leukocytes accumulate within adjacent vessels.

C. perfringens alpha-toxin is the major virulence factor in gas gangrene with inflammatory myopathies (Williamson and Titball 1993, Awad et al. 1995). The toxin, which exhibits phospholipase C (PLC) and sphingomyelinase activities, causes hemolysis, necrosis, and death, and the activation of neutrophils and release of cytokines (Sakurai, Nagahama and Oda 2004). Bryant reported that the intramuscular injection of alpha-toxin caused a rapid and irreversible decline in skeletal muscle blood flow due to toxin-induced intravascular aggregates of plates, leukocytes and fibrin (Bryant et al. 2000a, Bryant et al. 2000b). Neutrophils in these aggregates often bordered the endothelium but all remained intravascular (Bryant et al. 2000a). These findings suggested that the large heterotypic aggregates of platelets and leukocytes generated by alpha-toxin also contributed to impairment of the tissue inflammatory response. We have reported that alpha-toxin-induced activation of endogenous PLC and sphingomyelinase via a pertussis toxin (PT)-sensitive GTP-binding protein (Gi) plays an important role in the hemolysis of rabbit and sheep erythrocytes, respectively (Ochi et al. 1996, Ochi et al. 2004, Oda et al. 2008).

Recently, we revealed that the tyrosine kinase A (TrkA) receptor plays an important role in the release of superoxides and cytokines (Oda et al. 2006, Oda et al. 2008). This review will present findings about the signal transduction via TrkA receptor induced by alpha-toxin and summarize information about its likely role in inflammatory disease, especially septic shock.

2. Role of TrkA on a inflammation induced by alpha-toxin

2.1. Signal transduction via TrkA receptor

The TrkA receptor is a 140-kDa transmembrane protein encoded by a proto-oncogene located on chromosome 1 (Martin-Zanca, Hughes and Barbacid 1986). The family of Trk receptor tyrosine kinases consists of TrkA, TrkB and TrkC. While these family members have highly conserved sequences, they are activated by different neurotrophins: TrkA by nerve growth factor (NGF), TrkB by Brain-derived neurotrophic factor (BDNF) or neurotrophin 4 (NT4), and TrkC by NT3. TrkA regulates proliferation and is important for development and maturation of the nervous system (Pierotti and Greco 2006). This receptor comprises a tyrosine-kinase domain in its intra-cytoplasmic region and five extracellular domains, including two immunoglobulin-like domains involved in NGF binding and responsible for the specific selectivity to bind NGF (Wiesmann et al. 1999). In humans, the TrkA receptor is expressed on cells throughout the nervous system (Muragaki et al. 1995) as well as on structural cells and other non-neuronal cells in the immune and neuroendocrine systems (Levi-Montalcini et al. 1995, Aloe et al. 1997, Bonini et al. 2002, Levi-Montalcini 1987). When NGF binds to the TrkA receptor, it induces receptor homodimerization, which initiates kinase activation and transphosphorylation (Kaplan et al. 1991). This kinase activation involves small G proteins (Ras, Rac, Rap-1), PLCγ, protein kinase C (PKC) and phosphatidylinositol-3 kinase (PI3K) in neural cells (Obermeier et al. 1993b, Obermeier et al. 1993a, Melamed et al. 1999, York et al. 2000, Wu, Lai and Mobley 2001). Phosphorylation at Tyr490 is required for association with Shc and activation of the Ras-MAP kinase cascade. Residues Tyr674/675 lie within the catalytic domain, and phosphorylation at this site reflects TrkA kinase activity (Segal and Greenberg 1996, Stephens et al. 1994, Obermeier et al. 1993a, Obermeier et al. 1993b, Yao and Cooper 1995). Point mutations, deletions and chromosomal rearrangements (chimeras) cause ligand-independent receptor dimerization and activation of TrkA.

The mitogen-activated protein kinase (MAPK) pathways are activated next: extracellular-regulated protein kinase (ERK) by the small G proteins; ERK, p38 and JUN-N-terminal kinase (JNK) MAPK by PKC; and p38 and JNK by PI3K (Kaplan and Miller 1997). PI3K in turn induces activation of protein kinase B (PKB or Akt) and PKCξ (York et al. 2000)(Fig. 1).

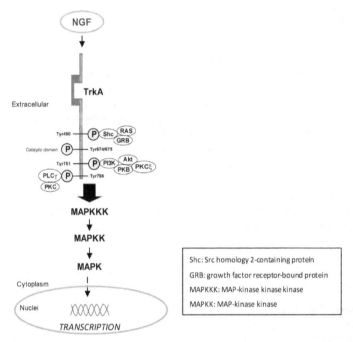

Figure 1. Signal transduction pathways of the TrkA receptor

2.2. Mechanism for the superoxide generation induced by alpha-toxin

The generation of superoxide in neutrophils has been reported to be stimulated by zymosan, 12-O-tetradecanoylphorbol 13-acetate (TPA), Ca^{2+} ionophores, and bacterial chemotactic peptides (Babior 1999). The signal transduction process leading to the stimulation has been studied extensively using N-formyl-methionyl-leucyl-phenylalanine (fMLP) (Kusunoki et al. 1992), platelet-activating factor (Yasaka, Boxer and Baehner 1982), and TPA (Nick et al. 1997, Pongracz and Lord 1998). It has been reported that these stimuli activated MAPK or PI3K in neutrophils (Shenoy, Gleich and Thomas 2003, Yamamori et al. 2004). Furthermore, these studies have demonstrated that the interaction of the ligands with receptors on neutrophils activates endogenous PLC with the formation of diacylglycerol (DG), which activates PKC, and inositol 1, 4, 5-trisphosphate (IP_3), inducing the release of Ca^{2+} from the endoreticulum, and that these products act synergistically to generate superoxide. Several studies also reported that phosphorylation of tyrosine kinases and activation of phospholipase D (PLD) were closely related to the generation of superoxide in neutrophils stimulated with agonists

(Garland 1992, Mitsuyama, Takeshige and Minakami 1993) and that activation of PLD resulted in the formation of PA, which was linked to the activation of NADPH oxidase (Bellavite et al. 1988, Olson, Tyagi and Lambeth 1990). We revealed that alpha-toxin-induced generation of superoxide is closely related to the activation of endogenous PKCθ via a combination of two events: production of DG on activation of PLC through a PT-sensitive GTP-binding protein and activation of phosphatidylinositide kinase 1 (PDK1) through the TrkA receptor (Oda et al. 2006).

There are three classes of PKC isotypes: classical PKC isotypes (PKCα, -β, and -γ) which have a C1 and C2 domain, bind DG, 1-oleoyl-2-acetyl-3-phosphoglycerol (OAG) and TPA, and are regulated by DG and Ca^{2+}; novel PKC isotypes (PKCδ, -ε, -η, and -θ), which have a C1 domain and novel C2 domain and are regulated by DG but not Ca^{2+}; and atypical isotypes (ζ/λ), which do not bind DG and are not regulated by these classical ligands (Le Good et al. 1998). Alpha-toxin induced phosphorylation of PKCθ and PKCζ/λ, and the generation of superoxide induced by the toxin was inhibited by rottlerin and calphostin C, an inhibitor of PKCθ. We reported that the formation of DG induced by alpha-toxin in rabbit neutrophils plays an important role in the generation of superoxide (Ochi et al. 2002). It therefore appears that the toxin-induced generation of superoxide is dependent on the activation of PKCθ, through binding of PKCθ phosphorylated by PDK1 to DG (Parekh, Ziegler and Parker 2000, Toker and Newton 2000). PKCθ has been reported to play an important role in activation of the protein 1 and NF-κB signaling pathway in T cells, production of interleukin-2, and apoptosis (Altman, Isakov and Baier 2000, Fan et al. 2004, Villalba et al. 1999, Villunger et al. 1999). Our data may provide clues to the role of PKCθ in neutrophils.

We reported that the alpha-toxin-stimulated generation of superoxide was related to the formation of DG through activation of endogenous PLC by a PT-sensitive GTP-binding protein in rabbit neutrophils (Ochi et al. 2002). U73122, an inhibitor of endogenous PLC, blocked the toxin-induced generation of superoxide and formation of DG in the cells, supporting that the toxin-induced increase in superoxide is dependent on the formation of DG by endogenous PLC. However, when the level of OAG incorporated into the cells was the same as the level of DG in the cells treated with 25 nM of the toxin, the level of OAG did not induce superoxide generation in the absence of the toxin but did in the presence of a near threshold dose (2.5 nM) of the toxin which did not induce production of DG. The result shows that the toxin-induced production of superoxide requires not only the formation of DG, but also the activation of other events.

It has been reported that the PI3K signaling pathway has an important role in several effector functions including the generation of superoxide (Yamamori et al. 2004). PI3K is known to generate phosphatidylinositol 3, 4, 5-trisphosphate (PIP₃), which is recognized by a pleckstrin homology domain identified as a specialized lipid-binding module (Le Good et al. 1998). Several papers have reported that PDK1 requires PIP₃ as its activator for effective catalytic activity (Le Good et al. 1998). Le Good et al. reported that there is a cascade involving PI3K, PDK1, and various members of the PKC superfamily in signal transduction (Le Good et al. 1998). Furthermore, the function of PKC family members is reported to depend on the phosphorylation of an activation loop by PDK1 (Le Good et al. 1998). LY294002 and wortmannin, both PI3K inhibitors, inhibited alpha-toxin-induced generation of superoxide and phosphorylation of PDK1 but did not affect the toxin-induced formation of DG. The result shows that the toxin-induced activation of PI3K occurs upstream of the phosphorylation of PDK1, which is an important step in the toxin-induced generation of superoxide. It is likely that the toxin-induced phosphorylation of PDK1 is a process independent of the toxin-induced formation of DG.

Tyrosine phosphorylation is thought to be crucial to the regulation of effector functions in neutrophils (Rollet et al. 1994). It is known that stimuli that induce tyrosine kinase activity in cells evoke the generation of PIP_1, PIP_2, and PIP_3. This tyrosine kinase activity is linked to the NGF receptors with intrinsic tyrosine kinase activity. Kannan et al. reported that NGF enhances the generation of superoxide induced by TPA in murine neutrophils (Kannan et al. 1991). Ehrhard et al. reported that human monocytes express the *trk* proto-oncogene, encoding the signal-transducing receptor unit for NGF, and that the interaction of NGF with monocytes triggers respiratory burst activity (Ehrhard et al. 1993). NGF, which did not induce the generation of superoxide in rabbit neutrophils, potentiated the events triggered by the toxin and caused superoxide to form in the presence of OAG, suggesting that a combination of the production of DG and stimulation of the NGF receptor induces severe activity in the generation of superoxide. The TrkA receptor was detected in rabbit neutrophils and found to be phosphorylated when the cells were treated with the toxin. Furthermore, immunoprecipitation using the anti-TrkA receptor antibody revealed direct binding of the toxin to the TrkA receptor. In addition, the antibody inhibited the toxin-induced generation of superoxide. These observations indicate that the interaction of alpha-toxin with TrkA receptors is important to the production of superoxide. In rabbit neutrophils, K252a, a TrkA inhibitor, and LY294002 inhibited the toxin-induced generation of superoxide and phosphorylation of PDK1 within specific concentration ranges, but PP2, a Src inhibitor, and AG1478, a epidermal growth factor receptor inhibitor, did not, supporting the finding that the TrkA receptor is involved in the toxin-induced increase in superoxide. The results obtained with the anti-TrkA antibody, LY294002, and K252a show that the activation of PI3K through direct binding of the toxin to the TrkA receptor results in production of PIP_3, which activates PDK1. In addition, PT inhibited the alpha-toxin-induced generation of superoxide and formation of DG, but not phosphorylation of PDK1, suggesting that a PT-sensitive GTP-binding protein plays a crucial role in the coupling to endogenous PLC, but not phosphorylation of PDK1. These observations indicate that the toxin independently induces activation of both endogenous PLC via a PT-sensitive GTP-binding protein and PDK1 via the TrkA receptor.

NGF, which binds to the TrkA receptor, is reported to be required for the differentiation and survival of sympathetic and some sensory and cholinergic neuronal populations (Howe et al. 2001). Furthermore, it has been reported that NGF is involved in inflammatory responses, an increase in mast cells in neonatal rats (Woolf et al. 1996), the degranulation of rat peritoneal mast cells (Woolf et al. 1996), and the differentiation of specific granulocytes (Kannan et al. 1991). The injection of *C. perfringens* cells or alpha-toxin into tissues is known to cause inflammation. Therefore, it is possible that the activation of the TrkA receptor by alpha-toxin is related to inflammation caused by C. *perfringens* in humans and animals.

H148G induced phosphorylation of PKCθ, but not production of DG, suggesting that the enzymatic activity of the toxin is essential for activation of endogenous PLC, but not activation of the TrkA receptor. It has been reported that binding of the C-domain, which does not contain the enzymatic site, to erythrocytes is important for the hemolysis induced by the toxin (Nagahama et al. 2002). It therefore is possible that the C-domain, the binding domain of alpha-toxin, plays a role in the binding of the toxin to the TrkA receptor and in the activation of signal transduction via the TrkA receptor.

Several studies have reported that the activation of PKC by various stimuli results in the generation of superoxide via the activation of MAPK systems (Coxon et al. 2003, Dewas et al. 2000, McLeish et al. 1998, Zu et al. 1998). K252a and U73122 inhibited the toxin-induced phosphorylation of PKCθ and ERK1/2 and generation of superoxide, suggesting that the toxin-induced production of superoxide is linked to the stimulation of the MAPK system via the activation of PKCθ. The toxin causes phosphorylation of ERK1/2, but not p38 and SAPK/JNK, implying that the process is dependent on a MAPK system containing MEK1/2 and MAPK/ERK1/2, but not systems containing p38 and SAPK/JNK.

It has been reported that PA directly or indirectly activated NADPH oxidase in a cell-free system of neutrophils (Erickson et al. 1999) and that PKCδ regulates phosphorylation of p67[phox] in human monocytes (Zhao et al. 2005). PKC also has been reported to activate directly NADPH oxidase (Johnson et al. 1998). However, PD98059 almost completely inhibited the toxin-induced production of superoxide near the inhibitory threshold dose of the inhibitor. Thus, it is unlikely that PA and PKC directly activate NADPH oxidase under the conditions used here.

We have shown that alpha-toxin induces formation of DG through the activation of endogenous PLC by a PT-sensitive GTP-binding protein and phosphorylation of PDK1 via stimulation of the TrkA receptor, so that DG and PDK1 synergistically activate PKCθ, and that the activation of PKCθ stimulates generation of superoxide through MAPK-associated signaling events in rabbit neutrophils (Fig. 2).

2.3. Mechanism for the cytokine release induced by alpha-toxin

Cytokines are immunoregulatory peptides with a potent inflammatory action, mediating the immune/metabolic response to an external noxious stimulus and fueling the transition from sepsis to septic shock, multiple organ dysfunction syndromes, and/or multiple organ failure (Tracey et al. 1987, Dinarello 2004, Riedemann, Guo and Ward 2003). It is thought that synergistic interactions between cytokines can cause or attenuate tissue injury (Calandra, Bochud and Heumann 2002). TNF-α, which is released early from neutrophils and macrophages, is one of the important cytokines involved in the pathophysiology of sepsis (Tracey et al. 1987, Lum et al. 1999). TNF-α-induced tissue injury is largely mediated through neutrophils, that respond by producing elastase, superoxide ion, hydrogen

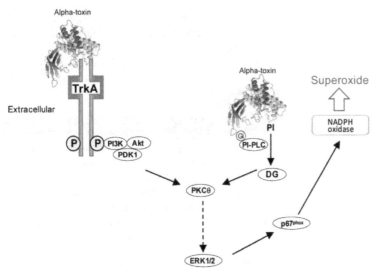

Figure 2. Signaling events involved in alpha-toxin-activated generation of superoxide

peroxide, sPLA2, PAF, leukotriene B1, and thromboxane A2 (Aldridge 2002). IL-1 stimulates the synthesis and release of prostagrandins, elastases, and collagenases and transendothelial microvascular cells, which respond by releasing the powerful neutrophil-stimulating agents, PAF and IL-8 (Leirisalo-Repo 1994). IL-1 and TNF-α are synergistic and share many biological effects in sepsis (Herbertson et al. 1995).

Anti-TNF-α antibody inhibited the death of mice induced by alpha-toxin. Furthermore, TNF-α-deficient mice were resistant to alpha-toxin. These observations suggest that the lethal effect of alpha-toxin is closely related to the release of TNF-α into the bloodstream. Stevens et al. and Bunting et al. suggested that alpha-toxin contributes indirectly to shock by stimulating production of endogenous mediators such as TNF-α and platelet-activating factor (Bunting et al. 1997, Stevens and Bryant 1997). It therefore appears that TNF-α released by alpha-toxin is important in enhancing the toxic actions of alpha-toxin in vivo. Consequently, inhibitors for release and expression of TNF-α may be worth pursuing as a novel therapeutic approach to the treatment of gas gangrene and sepsis caused by *C. perfringens*.

Cytokines such as the pro-inflammatory TNF-α, interleukin-1β (IL-1β) or transforming growth factor-β (TGF-β), increase the synthesis of NGF in airway structural cells. This stimulation has been evidenced in vitro in human pulmonary fibroblasts (Olgart and Frossard 2001, Micera et al. 2001), A549 epithelial cells (Pons et al. 2001) and bronchial smooth muscle cells (Freund et al. 2002). Studies also show that pro-inflammatory cytokines can act in concert to stimulate additional NGF secretion: TNF-α, for example, increases the secretion of NGF induced by IL-1β and interferon γ (IFN-γ) in fibroblasts (Hattori et al. 1994) and by interleukin-4 (IL-4) in astrocytes (Brodie et al. 1998). NGF synthesis in inflammatory conditions has also been demonstrated in vivo: elevated NGF concentrations are observed in cutaneous inflammation (Safieh-Garabedian et al. 1995) and in asthmatic airways (Olgart and Frossard 2001, Kassel, da Silva and Frossard 2001, Virchow et al. 1998). Taken together, these results suggest that pro-inflammatory cytokines, which are present at high levels in the airways of patients with asthma (Tillie-Leblond et al. 1999), might contribute to the elevated levels of NGF synthesis.

Corticosteroids are well known for their anti-inflammatory properties, particularly in asthmatic airways. Numerous studies report that the glucocorticoids dexamethasone and budesonide affect NGF expression. They cause a significant reduction in the increased NGF expression induced by pro-inflammatory cytokines; in one study, this action was shown to result from the repression of NGF gene transcription in endoneural fibroblasts from the rat sciatic nerve (Lindholm et al. 1990). Olgart and Frossard have reported that glucocorticoid treatment decreases the NGF secretion that the pro-inflammatory cytokines IL-1β and TNF-α stimulate in cultures of human pulmonary fibroblasts (Olgart and Frossard 2001) and in A549 epithelial cells (Pons et al. 2001).

These results suggested that the initial release of pro-inflammatory cytokines induced by alpha-toxin in vivo leads to the production of NGF, and the NGF released synergistically causes systemic inflammation such as sepsis and shock via activation of the TrkA receptor (Fig. 3).

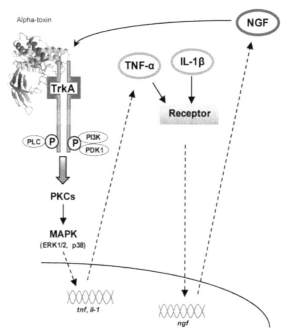

Figure 3. Alpha-toxin-induced release of pro-inflammatory cytokines and NGF

3. Conclusion

C. perfringens alpha-toxin, the main agent involved in the development of gas gangrene and septicemia, induces death, hemolysis, and the activation of macrophages and neutrophils. The toxin activated the MAPK-associated signal transduction from phospholipid metabolism and phosphorylation of TrkA. Penicillin is known to be highly effective in preventing the growth of microorganisms. In conclusion, treatment with TrkA inhibitors (tyrosine kinase inhibitors) and high doses of penicillin would be effective against diseases caused by C. perfringens.

Author details

Masataka Oda, Masahiro Nagahama, Keiko Kobayashi and Jun Sakurai
Faculty of Pharmaceutical Sciences, Tokushima Bunri University, Japan

4. References

Aldridge, A. J. (2002) Role of the neutrophil in septic shock and the adult respiratory distress syndrome. Eur J Surg, 168, 204-14.

Aloe, L., L. Bracci-Laudiero, S. Bonini & L. Manni (1997) The expanding role of nerve growth factor: from neurotrophic activity to immunologic diseases. Allergy, 52, 883-94.

Altman, A., N. Isakov & G. Baier (2000) Protein kinase Ctheta: a new essential superstar on the T-cell stage. Immunol Today, 21, 567-73.

Awad, M. M., A. E. Bryant, D. L. Stevens & J. I. Rood (1995) Virulence studies on chromosomal alpha-toxin and theta-toxin mutants constructed by allelic exchange provide genetic evidence for the essential role of alpha-toxin in Clostridium perfringens-mediated gas gangrene. Mol Microbiol, 15, 191-202.

Babior, B. M. (1999) NADPH oxidase: an update. Blood, 93, 1464-76.

Bellavite, P., F. Corso, S. Dusi, M. Grzeskowiak, V. Della-Bianca & F. Rossi (1988) Activation of NADPH-dependent superoxide production in plasma membrane extracts of pig neutrophils by phosphatidic acid. J Biol Chem, 263, 8210-4.

Bonini, S., A. Lambiase, G. Lapucci, F. Properzi, M. Bresciani, M. L. Bracci Laudiero, M. J. Mancini, A. Procoli, A. Micera, G. Sacerdoti, F. Levi-Schaffer, G. Rasi & L. Aloe (2002) Nerve growth factor and asthma. Allergy, 57 Suppl 72, 13-5.

Brodie, C., N. Goldreich, T. Haiman & G. Kazimirsky (1998) Functional IL-4 receptors on mouse astrocytes: IL-4 inhibits astrocyte activation and induces NGF secretion. J Neuroimmunol, 81, 20-30.

Bryant, A. E., R. Y. Chen, Y. Nagata, Y. Wang, C. H. Lee, S. Finegold, P. H. Guth & D. L. Stevens (2000a) Clostridial gas gangrene. I. Cellular and molecular mechanisms of microvascular dysfunction induced by exotoxins of Clostridium perfringens. J Infect Dis, 182, 799-807.

Bryant, A. E., R. Y. Chen, Y. Nagata, Y. Wang, C. H. Lee, S. Finegold, P. H. Guth & D. L. Stevens (2000b) Clostridial gas gangrene. II. Phospholipase C-induced activation of platelet gpIIbIIIa mediates vascular occlusion and myonecrosis in *Clostridium perfringens* gas gangrene. *J Infect Dis*, 182, 808-15.

Bunting, M., D. E. Lorant, A. E. Bryant, G. A. Zimmerman, T. M. McIntyre, D. L. Stevens & S. M. Prescott (1997) Alpha toxin from *Clostridium perfringens* induces proinflammatory changes in endothelial cells. *J Clin Invest*, 100, 565-74.

Calandra, T., P. Y. Bochud & D. Heumann (2002) Cytokines in septic shock. *Curr Clin Top Infect Dis*, 22, 1-23.

Coxon, P. Y., M. J. Rane, S. Uriarte, D. W. Powell, S. Singh, W. Butt, Q. Chen & K. R. McLeish (2003) MAPK-activated protein kinase-2 participates in p38 MAPK-dependent and ERK-dependent functions in human neutrophils. *Cell Signal*, 15, 993-1001.

Dewas, C., M. Fay, M. A. Gougerot-Pocidalo & J. El-Benna (2000) The mitogen-activated protein kinase extracellular signal-regulated kinase 1/2 pathway is involved in formyl-methionyl-leucyl-phenylalanine-induced p47phox phosphorylation in human neutrophils. *J Immunol*, 165, 5238-44.

Dinarello, C. A. (2004) Therapeutic strategies to reduce IL-1 activity in treating local and systemic inflammation. *Curr Opin Pharmacol*, 4, 378-85.

Ehrhard, P. B., U. Ganter, A. Stalder, J. Bauer & U. Otten (1993) Expression of functional trk protooncogene in human monocytes. *Proc Natl Acad Sci U S A*, 90, 5423-7.

Erickson, R. W., P. Langel-Peveri, A. E. Traynor-Kaplan, P. G. Heyworth & J. T. Curnutte (1999) Activation of human neutrophil NADPH oxidase by phosphatidic acid or diacylglycerol in a cell-free system. Activity of diacylglycerol is dependent on its conversion to phosphatidic acid. *J Biol Chem*, 274, 22243-50.

Fan, Y. Y., L. H. Ly, R. Barhoumi, D. N. McMurray & R. S. Chapkin (2004) Dietary docosahexaenoic acid suppresses T cell protein kinase C theta lipid raft recruitment and IL-2 production. *J Immunol*, 173, 6151-60.

Freund, V., F. Pons, V. Joly, E. Mathieu, N. Martinet & N. Frossard (2002) Upregulation of nerve growth factor expression by human airway smooth muscle cells in inflammatory conditions. *Eur Respir J*, 20, 458-63.

Garland, L. G. (1992) New pathways of phagocyte activation: the coupling of receptor-linked phospholipase D and the role of tyrosine kinase in primed neutrophils. *FEMS Microbiol Immunol*, 5, 229-37.

Hattori, A., S. Iwasaki, K. Murase, M. Tsujimoto, M. Sato, K. Hayashi & M. Kohno (1994) Tumor necrosis factor is markedly synergistic with interleukin 1 and interferon-gamma in stimulating the production of nerve growth factor in fibroblasts. *FEBS Lett*, 340, 177-80.

Herbertson, M. J., H. A. Werner, C. M. Goddard, J. A. Russell, A. Wheeler, R. Coxon & K. R. Walley (1995) Anti-tumor necrosis factor-alpha prevents decreased ventricular contractility in endotoxemic pigs. *Am J Respir Crit Care Med*, 152, 480-8.

Howe, C. L., J. S. Valletta, A. S. Rusnak & W. C. Mobley (2001) NGF signaling from clathrin-coated vesicles: evidence that signaling endosomes serve as a platform for the Ras-MAPK pathway. *Neuron*, 32, 801-14.

Johnson, J. L., J. W. Park, J. E. Benna, L. P. Faust, O. Inanami & B. M. Babior (1998) Activation of p47(PHOX), a cytosolic subunit of the leukocyte NADPH oxidase.

Phosphorylation of ser-359 or ser-370 precedes phosphorylation at other sites and is required for activity. *J Biol Chem*, 273, 35147-52.

Kannan, Y., H. Ushio, H. Koyama, M. Okada, M. Oikawa, T. Yoshihara, M. Kaneko & H. Matsuda (1991) 2.5S nerve growth factor enhances survival, phagocytosis, and superoxide production of murine neutrophils. *Blood*, 77, 1320-5.

Kaplan, D. R., B. L. Hempstead, D. Martin-Zanca, M. V. Chao & L. F. Parada (1991) The trk proto-oncogene product: a signal transducing receptor for nerve growth factor. *Science*, 252, 554-8.

Kaplan, D. R. & F. D. Miller (1997) Signal transduction by the neurotrophin receptors. *Curr Opin Cell Biol*, 9, 213-21.

Kassel, O., C. da Silva & N. Frossard (2001) The stem cell factor, its properties and potential role in the airways. *Pulm Pharmacol Ther*, 14, 277-88.

Kusunoki, T., H. Higashi, S. Hosoi, D. Hata, K. Sugie, M. Mayumi & H. Mikawa (1992) Tyrosine phosphorylation and its possible role in superoxide production by human neutrophils stimulated with FMLP and IgG. *Biochem Biophys Res Commun*, 183, 789-96.

Le Good, J. A., W. H. Ziegler, D. B. Parekh, D. R. Alessi, P. Cohen & P. J. Parker (1998) Protein kinase C isotypes controlled by phosphoinositide 3-kinase through the protein kinase PDK1. *Science*, 281, 2042-5.

Leirisalo-Repo, M. (1994) The present knowledge of the inflammatory process and the inflammatory mediators. *Pharmacol Toxicol*, 75 Suppl 2, 1-3.

Levi-Montalcini, R. (1987) The nerve growth factor 35 years later. *Science*, 237, 1154-62.

Levi-Montalcini, R., R. Dal Toso, F. della Valle, S. D. Skaper & A. Leon (1995) Update of the NGF saga. *J Neurol Sci*, 130, 119-27.

Lindholm, D., B. Hengerer, R. Heumann, P. Carroll & H. Thoenen (1990) Glucocorticoid Hormones Negatively Regulate Nerve Growth Factor Expression In Vivo and in Cultured Rat Fibroblasts. *Eur J Neurosci*, 2, 795-801.

Lum, L., B. R. Wong, R. Josien, J. D. Becherer, H. Erdjument-Bromage, J. Schlondorff, P. Tempst, Y. Choi & C. P. Blobel (1999) Evidence for a role of a tumor necrosis factor-alpha (TNF-alpha)-converting enzyme-like protease in shedding of TRANCE, a TNF family member involved in osteoclastogenesis and dendritic cell survival. *J Biol Chem*, 274, 13613-8.

Martin-Zanca, D., S. H. Hughes & M. Barbacid (1986) A human oncogene formed by the fusion of truncated tropomyosin and protein tyrosine kinase sequences. *Nature*, 319, 743-8.

McLeish, K. R., C. Knall, R. A. Ward, P. Gerwins, P. Y. Coxon, J. B. Klein & G. L. Johnson (1998) Activation of mitogen-activated protein kinase cascades during priming of human neutrophils by TNF-alpha and GM-CSF. *J Leukoc Biol*, 64, 537-45.

Melamed, I., H. Patel, C. Brodie & E. W. Gelfand (1999) Activation of Vav and Ras through the nerve growth factor and B cell receptors by different kinases. *Cell Immunol*, 191, 83-9.

Micera, A., E. Vigneti, D. Pickholtz, R. Reich, O. Pappo, S. Bonini, F. X. Maquart, L. Aloe & F. Levi-Schaffer (2001) Nerve growth factor displays stimulatory effects on human skin and lung fibroblasts, demonstrating a direct role for this factor in tissue repair. *Proc Natl Acad Sci U S A*, 98, 6162-7.

Mitsuyama, T., K. Takeshige & S. Minakami (1993) Tyrosine phosphorylation is involved in the respiratory burst of electropermeabilized human neutrophils at a step before diacylglycerol formation by phospholipase C. *FEBS Lett*, 322, 280-4.

Muragaki, Y., N. Timothy, S. Leight, B. L. Hempstead, M. V. Chao, J. Q. Trojanowski & V. M. Lee (1995) Expression of trk receptors in the developing and adult human central and peripheral nervous system. *J Comp Neurol*, 356, 387-97.

Nagahama, M., M. Mukai, S. Morimitsu, S. Ochi & J. Sakurai (2002) Role of the C-domain in the biological activities of *Clostridium perfringens* alpha-toxin. *Microbiol Immunol*, 46, 647-55.

Nick, J. A., N. J. Avdi, S. K. Young, C. Knall, P. Gerwins, G. L. Johnson & G. S. Worthen (1997) Common and distinct intracellular signaling pathways in human neutrophils utilized by platelet activating factor and FMLP. *J Clin Invest*, 99, 975-86.

Obermeier, A., H. Halfter, K. H. Wiesmuller, G. Jung, J. Schlessinger & A. Ullrich (1993a) Tyrosine 785 is a major determinant of Trk--substrate interaction. *EMBO J*, 12, 933-41.

Obermeier, A., R. Lammers, K. H. Wiesmuller, G. Jung, J. Schlessinger & A. Ullrich (1993b) Identification of Trk binding sites for SHC and phosphatidylinositol 3'-kinase and formation of a multimeric signaling complex. *J Biol Chem*, 268, 22963-6.

Ochi, S., K. Hashimoto, M. Nagahama & J. Sakurai (1996) Phospholipid metabolism induced by *Clostridium perfringens* alpha-toxin elicits a hot-cold type of hemolysis in rabbit erythrocytes. *Infect Immun*, 64, 3930-3.

Ochi, S., T. Miyawaki, H. Matsuda, M. Oda, M. Nagahama & J. Sakurai (2002) *Clostridium perfringens* alpha-toxin induces rabbit neutrophil adhesion. *Microbiology*, 148, 237-45.

Ochi, S., M. Oda, H. Matsuda, S. Ikari & J. Sakurai (2004) *Clostridium perfringens* alpha-toxin activates the sphingomyelin metabolism system in sheep erythrocytes. *J Biol Chem*, 279, 12181-9.

Oda, M., S. Ikari, T. Matsuno, Y. Morimune, M. Nagahama & J. Sakurai (2006) Signal transduction mechanism involved in *Clostridium perfringens* alpha-toxin-induced superoxide anion generation in rabbit neutrophils. *Infect Immun*, 74, 2876-86.

Oda, M., T. Matsuno, R. Shiihara, S. Ochi, R. Yamauchi, Y. Saito, H. Imagawa, M. Nagahama, M. Nishizawa & J. Sakurai (2008) The relationship between the metabolism of sphingomyelin species and the hemolysis of sheep erythrocytes induced by *Clostridium perfringens* alpha-toxin. *J Lipid Res*, 49, 1039-47.

Olgart, C. & N. Frossard (2001) Human lung fibroblasts secrete nerve growth factor: effect of inflammatory cytokines and glucocorticoids. *Eur Respir J*, 18, 115-21.

Olson, S. C., S. R. Tyagi & J. D. Lambeth (1990) Fluoride activates diradylglycerol and superoxide generation in human neutrophils via PLD/PA phosphohydrolase-dependent and -independent pathways. *FEBS Lett*, 272, 19-24.

Parekh, D. B., W. Ziegler & P. J. Parker (2000) Multiple pathways control protein kinase C phosphorylation. *EMBO J*, 19, 496-503.

Pierotti, M. A. & A. Greco (2006) Oncogenic rearrangements of the NTRK1/NGF receptor. *Cancer Lett*, 232, 90-8.

Pongracz, J. & J. M. Lord (1998) Superoxide production in human neutrophils: evidence for signal redundancy and the involvement of more than one PKC isoenzyme class. *Biochem Biophys Res Commun*, 247, 624-9.

Pons, F., V. Freund, H. Kuissu, E. Mathieu, C. Olgart & N. Frossard (2001) Nerve growth factor secretion by human lung epithelial A549 cells in pro- and anti-inflammatory conditions. *Eur J Pharmacol*, 428, 365-9.

Riedemann, N. C., R. F. Guo & P. A. Ward (2003) Novel strategies for the treatment of sepsis. *Nat Med*, 9, 517-24.

Rollet, E., A. C. Caon, C. J. Roberge, N. W. Liao, S. E. Malawista, S. R. McColl & P. H. Naccache (1994) Tyrosine phosphorylation in activated human neutrophils. Comparison of the effects of different classes of agonists and identification of the signaling pathways involved. *J Immunol*, 153, 353-63.

Safieh-Garabedian, B., S. Poole, A. Allchorne, J. Winter & C. J. Woolf (1995) Contribution of interleukin-1 beta to the inflammation-induced increase in nerve growth factor levels and inflammatory hyperalgesia. *Br J Pharmacol*, 115, 1265-75.

Sakurai, J., M. Nagahama & M. Oda (2004) *Clostridium perfringens* alpha-toxin: characterization and mode of action. *J Biochem*, 136, 569-74.

Segal, R. A. & M. E. Greenberg (1996) Intracellular signaling pathways activated by neurotrophic factors. *Annu Rev Neurosci*, 19, 463-89.

Shenoy, N. G., G. J. Gleich & L. L. Thomas (2003) Eosinophil major basic protein stimulates neutrophil superoxide production by a class IA phosphoinositide 3-kinase and protein kinase C-zeta-dependent pathway. *J Immunol*, 171, 3734-41.

Stephens, R. M., D. M. Loeb, T. D. Copeland, T. Pawson, L. A. Greene & D. R. Kaplan (1994) Trk receptors use redundant signal transduction pathways involving SHC and PLC-gamma 1 to mediate NGF responses. *Neuron*, 12, 691-705.

Stevens, D. L. & A. E. Bryant (1997) Pathogenesis of *Clostridium perfringens* infection: mechanisms and mediators of shock. *Clin Infect Dis*, 25 Suppl 2, S160-4.

Tillie-Leblond, I., J. Pugin, C. H. Marquette, C. Lamblin, F. Saulnier, A. Brichet, B. Wallaert, A. B. Tonnel & P. Gosset (1999) Balance between proinflammatory cytokines and their inhibitors in bronchial lavage from patients with status asthmaticus. *Am J Respir Crit Care Med*, 159, 487-94.

Toker, A. & A. C. Newton (2000) Cellular signaling: pivoting around PDK-1. *Cell*, 103, 185-8.

Tracey, K. J., Y. Fong, D. G. Hesse, K. R. Manogue, A. T. Lee, G. C. Kuo, S. F. Lowry & A. Cerami (1987) Anti-cachectin/TNF monoclonal antibodies prevent septic shock during lethal bacteraemia. *Nature*, 330, 662-4.

Villalba, M., S. Kasibhatla, L. Genestier, A. Mahboubi, D. R. Green & A. Altman (1999) Protein kinase ctheta cooperates with calcineurin to induce Fas ligand expression during activation-induced T cell death. *J Immunol*, 163, 5813-9.

Villunger, A., N. Ghaffari-Tabrizi, I. Tinhofer, N. Krumbock, B. Bauer, T. Schneider, S. Kasibhatla, R. Greil, G. Baier-Bitterlich, F. Uberall, D. R. Green & G. Baier (1999) Synergistic action of protein kinase C theta and calcineurin is sufficient for Fas ligand expression and induction of a crmA-sensitive apoptosis pathway in Jurkat T cells. *Eur J Immunol*, 29, 3549-61.

Virchow, J. C., P. Julius, M. Lommatzsch, W. Luttmann, H. Renz & A. Braun (1998) Neurotrophins are increased in bronchoalveolar lavage fluid after segmental allergen provocation. *Am J Respir Crit Care Med*, 158, 2002-5.

Wiesmann, C., M. H. Ultsch, S. H. Bass & A. M. de Vos (1999) Crystal structure of nerve growth factor in complex with the ligand-binding domain of the TrkA receptor. *Nature*, 401, 184-8.

Williamson, E. D. & R. W. Titball (1993) A genetically engineered vaccine against the alpha-toxin of *Clostridium perfringens* protects mice against experimental gas gangrene. *Vaccine*, 11, 1253-8.

Woolf, C. J., Q. P. Ma, A. Allchorne & S. Poole (1996) Peripheral cell types contributing to the hyperalgesic action of nerve growth factor in inflammation. *J Neurosci*, 16, 2716-23.

Wu, C., C. F. Lai & W. C. Mobley (2001) Nerve growth factor activates persistent Rap1 signaling in endosomes. *J Neurosci*, 21, 5406-16.

Yamamori, T., O. Inanami, H. Nagahata & M. Kuwabara (2004) Phosphoinositide 3-kinase regulates the phosphorylation of NADPH oxidase component p47(phox) by controlling cPKC/PKCdelta but not Akt. *Biochem Biophys Res Commun*, 316, 720-30.

Yao, R. & G. M. Cooper (1995) Requirement for phosphatidylinositol-3 kinase in the prevention of apoptosis by nerve growth factor. *Science*, 267, 2003-6.

Yasaka, T., L. A. Boxer & R. L. Baehner (1982) Monocyte aggregation and superoxide anion release in response to formyl-methionyl-leucyl-phenylalanine (FMLP) and platelet-activating factor (PAF). *J Immunol*, 128, 1939-44.

York, R. D., D. C. Molliver, S. S. Grewal, P. E. Stenberg, E. W. McCleskey & P. J. Stork (2000) Role of phosphoinositide 3-kinase and endocytosis in nerve growth factor-induced extracellular signal-regulated kinase activation via Ras and Rap1. *Mol Cell Biol*, 20, 8069-83.

Zhao, X., B. Xu, A. Bhattacharjee, C. M. Oldfield, F. B. Wientjes, G. M. Feldman & M. K. Cathcart (2005) Protein kinase Cdelta regulates p67phox phosphorylation in human monocytes. *J Leukoc Biol*, 77, 414-20.

Zu, Y. L., J. Qi, A. Gilchrist, G. A. Fernandez, D. Vazquez-Abad, D. L. Kreutzer, C. K. Huang & R. I. Sha'afi (1998) p38 mitogen-activated protein kinase activation is required for human neutrophil function triggered by TNF-alpha or FMLP stimulation. *J Immunol*, 160, 1982-9.

Chapter 3

Function of Flotillins in Receptor Tyrosine Kinase Signaling and Endocytosis: Role of Tyrosine Phosphorylation and Oligomerization

Nina Kurrle, Bincy John, Melanie Meister and Ritva Tikkanen

Additional information is available at the end of the chapter

1. Introduction

There are two homologous members of the flotillin family which have been designated as flotillin-1/reggie-2 and flotillin-2/reggie-1, owing to their almost simultaneous discovery by two separate research groups (1, 2). In 1997, these proteins were described under the name "reggies", since they were found to be upregulated during the regeneration of retinal ganglion cells of the goldfish, following an optic nerve lesion (2). In the same year, Bickel *et al.* independently discovered them as components of detergent resistant fractions, and due to their ability to float in density gradients ascribed them as flotillins (1).

Biological membranes are complex structures composed of numerous proteins and lipids. As shown in Figure 1A, flotillin-1 and flotillin-2 are both constitutively associated with sphingolipid and cholesterol enriched membrane microdomains, known as lipid rafts (1, 3-5). Especially flotillin-1 is often considered as a *bona fide* marker protein for lipid rafts. The lipid raft concept was first postulated in 1997 by Kai Simons and Elina Ikonen (6), and has since then been refined and reshaped throughout the years. Nowadays, lipid rafts can be described as fluctuating nanoscale membrane assemblies composed of proteins, sphingolipids and cholesterol, which can coalesce to form platforms that are important for cell signaling, viral infection and membrane trafficking (7). It has been proposed that proteins with a high affinity for an ordered lipid environment, often achieved by post-translational modifications, will preferentially associate with lipid rafts (8). We have shown that flotillin proteins are linked to the plasma membrane by means of fatty acid modifications, especially myristoylation and palmitoylation (4), and both modifications facilitate the interaction with membrane cholesterol (our unpublished results). Earlier studies proposed that flotillins are localized in caveolae-type of lipid rafts, but later findings demonstrated that flotillins rather reside outside of caveolae and are thought to define their own subtype of lipid rafts (9).

The first description of flotillin-2 actually took place in 1994, when the group of Madeleine Duvic cloned an N-terminally truncated version of flotillin-2 from human epidermal keratinocytes and named it epidermal surface antigen (ESA) (10). However, in later studies by the same group, it turned out that the true ESA (ECS-1 antigen) is actually distinct from flotillin-2 (11). The flotillin-2 cDNA originally cloned by Schroeder *et al.* represents a truncated protein with a lower molecular mass that has not been detected *in vivo* and most likely came about due to a missing 5' end of the cloned cDNA and false assignment of the ATG start codon. A monoclonal antibody ECS-1 raised against human keratinocytes was used for the identification of ESA. Curiously, a very similar approach with a monoclonal antibody against cell surface proteins was used by the Stuermer group to identify reggie-1/flotillin-2 and reggie-2/flotillin-1 as regeneration proteins of goldfish retinal ganglion cells (2). In this case, the true antigen of the M802 antibody was later shown to be identical with Thy-1, a glycophosphatidyl-inositol (GPI)-linked neuronal protein (12). Since flotillins have been shown not to traverse the membrane, it is unlikely that they were directly recognized by the antibodies used for the screenings. A more likely explanation is that flotillins were associated with the true antigens as molecular complexes and thus accidentally discovered. This is also supported by the data showing that flotillins are generally strongly associated with GPI-linked proteins (13-15). However, the details of their interaction have mechanistically not been clarified, since neither flotillins nor GPI-anchored proteins completely traverse the membrane. Thus, their interaction may be mediated by a yet to be identified transmembrane protein. Accordingly, we were able to show that both flotillins are coprecipitated after immunoprecipitation of surface biotinylated membrane proteins in confluent human epithelial MCF10A cells (our unpublished results).

Both flotillin-1 and flotillin-2 show a constitutive expression in almost all mammalian cell types and are highly conserved from fly to man (reviewed in 16). Furthermore, flotillin-like proteins are also present in bacteria (17), plants (18), fungi and metazoans but are absent in yeast and *Caenorhabditis elegans* (19). In humans, a 15 kb single-copy flotillin-1 gene with 13 exons is located on chromosome 6 (6p21.3), giving rise to a 1.9 kb mRNA transcript. The ORF encodes a 427 residue protein with a molecular mass of 47 kDa (20). A single copy of flotillin-2 gene is located on chromosome 17 (17q11-12) which encodes for a 48 kDa protein (10). Human flotillin-1 shares 98% identity with the murine protein and 47% identity with human flotillin-2. Flotillins are classified as part of the SPFH (Stomatin/ Prohibitin/ Flotillin/HflK/C) superfamily of proteins, also called prohibitin homology (PHB) proteins (21, 22). Different from other members of the SPFH superfamily, flotillins exhibit a conserved domain termed the flotillin domain C-terminally to the SPFH domain (Fig. 1B). This domain is characterized by several repeats of glutamic acid and alanine (EA repeats) and is predicted to form coiled coil structures (23, 24). A few years ago, the crystal structure of the core SPFH domain of mouse flotillin-2 was solved, and it shows partial similarity to the respective segment of *Pyrococcus horikoshii* stomatin (25). This structural region includes amino acids 43 to 173 of the SPFH domain of flotillin-2 and depicts that flotillin-2 SPFH domain has a compact, ellipsoid-globular structure, comprising four alpha helices and six beta strands. Although the structure of the flotillin domain has not yet been solved, it is predicted to show a high propensity for alpha helices. This could explain why flotillin domains are engaged in coiled coil structures that are important for the oligomerization of flotillins.

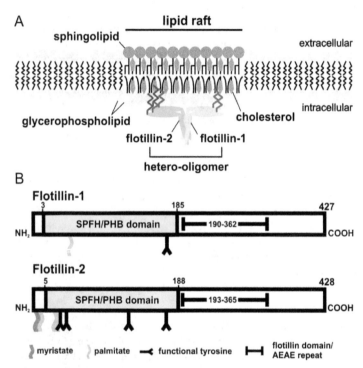

Figure 1. Flotillin-1 and flotillin-2 form oligomers in lipid rafts. A: Flotillin-1 and flotillin-2 are associated with specialized membrane microdomains, known as lipid rafts. Lipid rafts are defined as fluctuating nanoscale assemblies in membranes, composed of certain proteins and enriched in sphingolipids and cholesterol. They can coalesce to form platforms that are important for cell signaling, viral infection and membrane trafficking. Flotillin-1 and flotillin-2 are linked to the membrane by acylation, namely palmitoylation (both flotillins) and myristoylation (only flotillin-2). An important feature of flotillin function is the formation of homo- and hetero-oligomers. B: Flotillins contain in their N-terminal region a domain homologous with the prohibitin family of proteins, whereas their C-terminus is characterized by the unique flotillin domain with AEAE repeats.

2. Modifications, oligomerization and trafficking of flotillins

Flotillin-1 and flotillin-2 are ubiquitously expressed proteins and exhibit a relatively complementary tissue distribution. Northern blot analysis showed a high tissue-specific expression of flotillin-1 in murine brain, fat and heart tissue (1). High protein levels of flotillin-1 and flotillin-2 were found, e.g. in murine lung, thymus, bladder, cerebrum and hypothalamus tissue lysates (24, our unpublished results). Additionally, it has been shown that the expression of flotillin-2 increases during differentiation and myogenesis of C2C12 cells, whereas flotillin-1 increases during 3T3-L1 cell adipogenesis (1) and osteoclastogenesis (26). In contrast, during differentiation of rat pheochromocytoma PC12 cells, no change in the expression of flotillin-1 or flotillin-2 was observed (24).

The subcellular localization of flotillin-1 and flotillin-2 is as diverse as their tissue expression pattern and seems to be cell type specific. In glandular cervical cancer HeLa cells, PC12 cells, human hepatocellular carcinoma HepG2 cells, Chinese hamster ovary (CHO) cells and human breast cancer MCF-7 cells, flotillins mainly reside at the inner leaflet of the plasma membrane and to a minor extent in vesicular structures. A major localization in vesicular structures is found in astrocytes, the human HaCaT keratinocyte cell line and mouse embryonic fibroblast (MEF) cells. Colabeling experiments and immunogold electron microscopy studies identified the vesicular structures as endosomes and lysosomes in astrocytes, Jurkat, PC12 and HeLa cells (13, 14, 27-30). In line with the previous results, endogenous labeling of flotillin-2 in MCF10A cells shows a more prominent localization at the plasma membrane, whereas flotillin-1 is found in vesicular structures in non-confluent cells. However, with increasing confluency and advanced differentiation, flotillin-1 translocates to the plasma membrane (our unpublished results), a feature already observed in differentiating mouse 3T3-L1 cells (31). Rajendran *et al.* (32, 33) showed that flotillin-1 and flotillin-2 reside in preassembled structures under resting conditions in immature and mature hematopoietic cells, but they can move into uropods during cell migration and in immunological synapses during T-cell activation. These two studies highlight the fact that the subcellular localization of flotillin proteins is highly versatile and depends on the cell type and the differentiation status.

Total internal reflection fluorescence (TIRF) microscopy revealed that flotillin-1 and flotillin-2 exhibit a vesicular cycling at the plasma membrane (28), which seems to be independent of clathrin and caveolin (9, 13, 34, 35). Microtubule disruption in HeLa cells results in the accumulation of flotillin-1/flotillin-2 positive vesicles in the cytosol (28), while the same treatment in murine neuroblastoma N2a cells did not affect flotillin-2 localization (36). Manipulation of the actin cytoskeleton might also have a cell type dependent influence on the mobility of flotillins. In FRAP (fluorescence recovery after photobleaching) experiments, an increased lateral mobility of flotillin-2 after disruption of the actin cytoskeleton with cytochalasin D in N2a cells was observed, while treatment with the actin stabilizing compound jasplakinolide reduced the lateral mobility. On the contrary, the use of cytochalasin D in HeLa cells had no significant effect on flotillin-2 mobility (36). Recently, Affentrager *et al.* (37) showed that a dynamic actin turnover is required for capping of flotillin-1 and flotillin-2 in primary human peripheral blood T-lymphocytes, which underscores the importance of the actin cytoskeleton for flotillin localization and trafficking. The lateral motility is also dependent on the localization itself, since at least in Jurkat T cells, flotillin-2 is selectively immobilized in the region of the preformed cap and only shows a high lateral mobility at the plasma membrane outside the cap (29). In other studies, flotillin-

1 was found in exosomes of rat and human reticulocytes, human erythroleukemia K562 cells, lymphoid-B Daudi cells, human breast carcinoma MCF-7 cells and in exosomes of the cerebrospinal fluid (38-40). Flotillin-2 was detected in exosomes derived from MCF-7 cells and oligodendroglial Oli-neu cells (40, 41). Flotillin-1 was additionally found in mature phagosomes of murine J774 macrophages (42) and in the nucleus of human prostate cancer PC-3 cells (43).

Localization and trafficking of flotillins in the Golgi apparatus is still controversial. Nevertheless, evidence suggests that the involvement of Golgi in flotillin trafficking is also a cell type-specific process. In non-differentiated PC12, normal rat kidney (NRK), CHO and HeLa cells, flotillin-1 localizes to the Golgi complex (44-46). Also in rat brain tissue sections, flotillin-1 was found in the Golgi (47). In contrast, Morrow *et al.* (3) reported that flotillin-1 reaches the plasma membrane via a brefeldin A-resistant pathway in baby hamster kidney (BHK) cells. However, they did not rule out the possibility of Golgi involvement in flotillin-1 trafficking, as a GFP-tagged construct of the flotillin-1 SPFH domain colocalized with the cis-medial Golgi marker GM130 during early times of expression. Less is known about flotillin-2 and Golgi complex association. Only the study of Langhorst *et al.* (28) showed by immunogold EM microscopy that flotillin-2 was localized in vesicles in the Golgi field in Jurkat cells, but not in Golgi stacks.

Chemokines, growth factors and certain proteins can affect the localization and the trafficking of flotillin proteins. In Satb2-negative neuronal cells, flotillin-1 becomes more clustered in response to semaphorin 3a treatment, which acts as a potent guidance cue presented to cortical axons *in vivo* (48). In primary human peripheral blood T-lymphocytes, flotillin-1 and flotillin-2 are randomly distributed along the inner leaflet of the plasma membrane, and upon stimulation with stromal cell derived factor 1 (SDF-1), flotillins form stable caps in the uropod (32, 37). The best studied growth factor that has an effect on flotillin trafficking is the epidermal growth factor (EGF), as originally shown by us (30). Recently, we showed that flotillin-1 is indeed involved in the activation and downstream signaling of the EGF receptor (EGFR) (27). Upon stimulation with EGF, flotillin-2 becomes tyrosine phosphorylated at several tyrosine residues by Src kinases (30). This phosphorylation event is dependent on the activation of the EGFR, since the EGFR autophosphorylation inhibitor AG1478 prevents flotillin-2 phosphorylation. In regard to its localization, EGF stimulation results in a Tyr163-dependent translocation of flotillin-2 into the late endosomal compartment (30). Experiments in the group of B. Nichols confirmed our findings stating that Y163 is important for flotillin-2 internalization. In addition, it was shown that the corresponding tyrosine in flotillin-1, Y160, is crucial for flotillin-1 internalization (49). Other growth factors, such as hepatocyte growth factor (HGF) or insulin also affect the localization of flotillin proteins (Figure 2, and our unpublished results). Figure 2 shows that ten minutes after HGF stimulation, flotillin-1 is translocated from the plasma membrane to late endosomes/lysosomes where it colocalizes with LAMP3.

Posttranslational modifications of eukaryotic proteins are often important cues for localization, trafficking and membrane association of proteins (50). Two important modifications are N-myristoylation and S-palmitoylation. N-myristoylation is an irreversible

unstimulated   10 min HGF

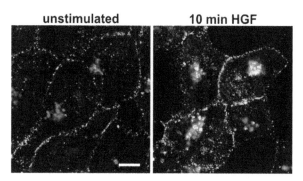

Figure 2. Flotillin-1 translocates to late endosomes upon 10 min of hepatocyte growth factor (HGF) stimulation. Staining of endogenous flotillin-1 (green) and endogenous lysosome-associated membrane glycoprotein 3 (LAMP-3; red) in HeLa cells stimulated for 10 minutes with HGF. Scale bar: 15 μm

acylation reaction at an N-terminal glycine residue, which is catalyzed by the enzyme N-myristoyl transferase (51). S-palmitoylation is a reversible reaction, in which the acyl group is transferred to cysteine residues by palmitoyl acyltransferases, referred to as the DHHC family (52, 53). In some proteins, e.g. Src-related protein tyrosine kinases, which are both myristoylated and palmitoylated, the myristoylation reaction is a prerequisite for the palmitoylation to occur (54). Our experiments with overexpressed flotillin-2 constructs in HeLa cells showed that flotillin-2 is myristoylated at Gly2 and palmitoylated at Cys4, Cys19 and Cys20, and Cys4 seems to be the major palmitoylation site (4). The myristoylation at Gly2 is essential for membrane targeting of flotillin-2, as the mutant flotillin-2 Gly2Ala remains fully soluble upon overexpression. Furthermore, the mutation of the cysteine residues to alanine and the concomitant loss of palmitoylation sites resulted in more soluble flotillin-2 mutants (4). In the study by Li *et al.* (53), the palmitoyl acyltransferase DHHC5 was identified as the enzyme to palmitoylate flotillin-2. In neuronal stem cells of DHHC5 gene-trapped mice, flotillin-2 palmitoylation was abolished. For flotillin-1, which is not myristoylated, it was shown that plasma membrane localization is dependent on palmitoylation of a conserved cysteine residue (Cys34) within its SPFH domain (3). This finding is supported by the study of Affentrager *et al.* (37), in which overexpressed flotillin-1 Cys34Ala exhibited a diffuse cytosolic localization in human T-lymphoblasts. In contrast, palmitoylation of this cysteine seems not to play a crucial role for plasma membrane targeting in adipocytes. Instead, deletion of the second hydrophobic stretch (residues 134 – 151) results in a similar cytoplasmic localization (31). This finding suggests that the effect of Cys34 palmitoylation in flotillin-1 function might be cell type specific.

Oligomeric proteins are prevalent in nature and comprise approximately one third of all known proteins. From the perspective of protein evolution, oligomerization may be of advantage for a functional control such as allosteric regulation (55). In the case of flotillin proteins, oligomerization is an essential feature for membrane raft association, endocytosis and EGF signaling (30, 56). Flotillin-1 and flotillin-2 form both homo- and hetero-oligomers (4, 5, 23, 34), and Solis *et al.* (23) suggested that flotillin oligomers exist in a monomer-tetramer equilibrium. The homo-oligomerization of flotillin-2 is mediated by the C-terminal half of the protein (amino acid residues 208-428), which seems to enhance the association of flotillin-2 with the plasma membrane and its membrane raft localization (4, 29). The C-terminal half of flotillin-1 and flotillin-2 contains the so called flotillin-domain, which contains several EAEA repeats and is predicted to form three adjacent coiled-coil structures. The formation of flotillin oligomers was shown to be dependent on the second and partially on the first coiled-coil structure (23). Using the flotillin-2 Y163A mutant, we showed that not the phosphorylation as such but hetero-oligomerization with flotillin-1 is indispensable for flotillin-2 endocytosis. In addition, our experimental data implicate that after EGF stimulation, preexisting flotillin oligomers form larger oligomeric complexes in which the stoichiometry of 1:1, also shown by Frick *et al.* (34), is preserved (56). In summary, these data show that not only the coiled coil structures within the C-terminal half of flotillins are important for the oligomerization, but also the tyrosine residue 163 in flotillin-2 plays a role in the hetero-oligomerization with flotillin-1.

Results of the Stuermer group have shown that ectopic expression of the presumed oligomerization domain of flotillin-2 comprising the amino acids 184 to 390 in Jurkat T cells interferes with function of flotillins in these cells and prevents the formation of macrodomains upon crosslinking of GM1 by cholera toxin (29). Hence, this fragment, termed R1EA, most likely prevents the proper oligomerisation of flotillins, thus functioning in a trans-negative manner. Upon overexpression of R1EA, insulin-like growth factor induced neurite outgrowth was abrogated in N2a neuroblastoma cells, and axon differentiation was impaired in hippocampal neurons (57). In addition, in R1EA expressing cells, recruitment of CAP/ponsin, which interacts with flotillin-1, to focal contacts was affected, resulting in imbalanced activation of Rho family GTPases Rac1 and Cdc42. Interestingly, focal adhesion kinase (FAK) activity was increased by R1EA expression. However, other signaling pathways such as ERK1/2, PKC, and PKB signals were unaltered (57). In line with this, previous studies have shown that upon growth factor stimulation of PC12 cells, protein kinase Pyk2, ubiquitin ligase Cbl and the CAP homolog ArgBP-2 are recruited to lipid rafts (58). Moreover, cholesterol depletion resulted in a diminished neurite outgrowth, suggesting involvement of membrane rafts (58). It has been proposed that ArgBP-2 associates with Cbl and Pyk2 via its SH3 domain and recruits them to lipid rafts by interacting with flotillin-1 through its SoHo domain (58). These findings point towards a vital role of flotillins in cytoskeletal remodeling and neurite growth at least in cultured cells. However, neither the flotillin-1 knockout mouse model (59) nor flotillin-2 deficient mice (our unpublished data) show a major neuronal phenotype. It thus remains to be clarified if flotillins are only required under conditions of neuronal regeneration or if they are generally important for the physiology of neurons in mammals.

3. In search of the molecular function of flotillins

As can be expected from such highly conserved and ubiquitous proteins, flotillins have been shown to play a role in a large number of vital cellular processes. Their original discovery as "reggies" suggested a function in neuronal regeneration in the goldfish (2, 45), which has been later verified by further studies from the same group (57, 60, 61). However, the regenerative capacity of the optic nerve is poorly conserved in mammals, making a generalization of the regenerative function of flotillins in the nervous system of mammals difficult. Recent studies have confirmed the relevance of flotillins for axon regeneration in zebrafish as well as in mammalian cell cultures. Apparently, downregulation of flotillins by flotillin specific morpholinos in the zebrafish resulted in a massive reduction in the number of regenerating axons (60). Since the role of flotillins in neuronal regeneration has recently been directly reviewed, we here omit a long discussion of the topic. Instead, the reader is referred to the recent review of C. Stuermer (62).

Flotillins also appear to be involved in membrane trafficking processes such as endocytosis and phagocytosis (13, 34, 42), and they have been suggested to be involved in the regulation the actin cytoskeleton (4, 11, 57). One emerging important function of flotillins, especially of flotillin-1, is the regulation of membrane receptor signaling, e.g. through mitogen activated protein kinases (MAPK) (27, 30, 43, 56, 63-67). Below, we have summarized the data on the function of flotillins in some of these processes.

Flotillins in endocytosis

During their discovery, flotillins were described to be associated with membrane rafts (1, 24). First findings suggested that flotillins would be integral membrane proteins of the cave-like invaginations called caveolae (1, 24), but these findings have later been disputed by several other studies (3, 9, 13, 34). However, the raft association of flotillins appears to be important for their function, and flotillin-1 has even been suggested to define a novel non-clathrin, non-caveolin endocytosis pathway that would operate via membrane rafts (13, 34). In eukaryotic cells, uptake of membrane bound molecules such as receptors can take place by means of several different pathways. The best characterized and the most common pathway for endocytosis is mediated by clathrin coated vesicles which are formed at the plasma membrane by the assembly of a coated pit which then forms a clathrin coated vesicle. However, recent research from several groups has revealed that endocytosis can be divided into numerous pathways, many of which do not require clathrin, and are thus designated as clathrin-independent. For a general introduction on endocytosis, the reader is referred to these excellent reviews (68-70). Some factors that regulate the clathrin independent pathways have been identified in the recent years, and it is clear that small GTPases such as Cdc42, Rho A or Arf (ADP ribosylation factor) family are involved, each determining their own endocytic pathway. The role of flotillins in endocytosis has recently been reviewed in detail (71), and we thus here only present a brief overview and a summary of the very recent studies not included in the review of Otto and Nichols.

Early after their discovery, flotillins were shown to associate and colocalize with GPI-anchored proteins. A role in the internalization of GPI-anchored proteins such as CD59 and of the glycosphingolipid GM1 has been demonstrated (13, 34, 72). Flotillins have also been suggested to be involved in the endocytosis of the GPI-linked Nodal coreceptor cripto (73). Although flotillins are affirmed regulators of cholera toxin subunit B endocytosis (13, 72), which is mediated by the binding of the toxin to GM1, no evidence for a direct role of

Function of Flotillins in Receptor Tyrosine Kinase Signaling and Endocytosis: Role of Tyrosine
Phosphorylation and Oligomerization

67

flotillins in the endocytosis of the Shiga toxin or the plant toxin ricin was found (46). However, flotillins were shown to control other trafficking steps of these toxins, most likely a retrograde transport step towards Golgi/ER. Intriguingly, depletion of flotillins resulted in a higher degree of mannosylation of ricin, and the toxicity of both ricin and Shiga toxin was increased (46). Thus, flotillins are important factors in controlling not only the endocytosis but also other steps of the cellular trafficking of various molecules.

In addition to GPI-anchored proteins, flotillins have recently been shown to participate in the protein kinase C (PKC) induced endocytosis of neurotransmitter receptors such as the dopamine transporter DAT (also designed as solute carrier family 6, member 3 or SLC6A3) and glial glutamate transporter, GLT1 (74). PKC was shown to phosphorylate flotillin-1 at a Ser residue, and this phosphorylation was required for the endocytosis of DAT (74). Recent publications have revealed a role for flotillins in the endocytosis of the Alzheimer amyloid precursor protein (APP), the amyloidogenic processing of which requires its endocytosis (75). In addition, Niemann-Pick C1-like 1 (NPC1L1), an important mediator of the uptake of dietary cholesterol, relies on flotillins for its endocytosis, and depletion of flotillins drastically reduces cholesterol uptake in cultured cells (76). Curiously, APP, NPC1L1 and DAT are all endocytosed by means of clathrin mediated endocytosis, whereas flotillins as raft associated proteins have been rather suggested to define a non-clathrin endocytosis pathway, and there is very little colocalization between clathrin and flotillins (13). Thus, it is possible that rather than directly mediating raft dependent endocytosis through flotillin microdomains, flotillins might exert their function by first facilitating the uptake of some cargo molecules into rafts and their efficient clustering, but the act of endocytosis may only take place after transfer of the cargo into clathrin coated pits. Hence, it will be important in the future to study the exact relationship of flotillins and clathrin dependent endocytosis.

We have recently shown that flotillins are involved in EGFR/MAP kinase signaling, and siRNA mediated downregulation of flotillin-1 results in severe impairment of both EGFR phosphorylation and MAPK signaling (27). Since flotillins also become phosphorylated upon EGFR signaling (30), it would not be surprising if they were involved in EGFR endocytosis, which has been shown to take place by means of both clathrin dependent and raft mediated endocytosis (77, 78). However, according to our recent findings, flotillin-1 depletion does not significantly affect the EGF-induced endocytosis of EGFR (27). Furthermore, EGFR is efficiently endocytosed even in cells depleted of both clathrin and flotillin-1 (our unpublished data). Interestingly, recent findings from the Stuermer group have implicated that depletion of flotillin-2/reggie-1 may affect the endocytosis of EGFR (79). This finding is somewhat counter intuitive as flotillin-2 depletion induces a significant depletion of flotillin-1 as well, and flotillin-2 knockdown cells express only minor amounts of both flotillins. Flotillin-1 knockdown cells in turn still exhibit substantial flotillin-2 expression. Thus, these results imply that the flotillin-2 homo-oligomers that are still present in flotillin-1 knockdown cells may be capable of supporting normal EGFR endocytosis, whereas in the absence of flotillins, EGFR uptake from the plasma membrane is impaired. However, one also has to keep in mind some important methodological differences between our study (27) and that of Solis et al. (79). Our EGF uptake assays were performed with surface-bound iodinated EGF using short endocytosis times (up to 6 min), which allows the calculation of endocytosis rate constants for the receptor, providing a reliable and truly quantitative measurement of the early endocytosis and uptake of the receptor from the plasma membrane (80). We observed no difference between flotillin-1 knockdown and control cells even after very short endocytosis times (2, 4 or 6 min). However, Solis *et al.* used methods such as continuous uptake of fluorescent EGF for 5-10 min, or endocytosis of biotinylated EGFR for 120 min. In the latter assay, they observed an increased amount of EGFR remaining at the cell surface of flotillin-2 knockdown cells. However, also the total EGFR was increased in flotillin-2 depleted cells after 120 min EGF stimulation. Since substantial degradation of EGFR (and EGF) will evidently take place over such a long time, these results might indicate that the degradation and not endocytosis of EGFR is impaired, resulting in generally higher amount of remaining EGFR in flotillin-2 knockdown cells. Furthermore, when longer time points are used (10 - 120 min), some of the receptor will also be recycled back to the plasma membrane, thus biasing these results. In fact, considerable recycling of EGFR has been shown to take place in A431 cells (81). Hence, an increased amount of EGFR on the surface of flotillin-2 knockdown cells could also be a result of increased recycling combined with a reduced degradation. In addition to this, the A431 cells used by Solis *et al.* massively overexpress EGFR and are thus widely considered as a poor model system for EGFR trafficking, since the high amounts of EGFR may result in saturation of the endocytosis pathways normally used by the receptor and redirection of the receptor to pathways that otherwise play a negligible role. Therefore, the role of each flotillin in EGFR endocytosis should be reevaluated and directly compared in the same cell system with methods that are suitable for unbiased measurement of endocytosis, such as iodinated EGF uptake.

One important open question on the role of flotillins in endocytosis is if the flotillin-mediated endocytosis of various cargo molecules is dependent on the GTPase dynamin or not, since the studies have resulted in contradictory results, depending on the cell system used and cargo molecules studied. Since flotillins are endocytosed from the plasma membrane as a result of EGF stimulation (30) but appear not to directly affect EGFR endocytosis (27), they may in some cases also themselves represent cargo molecules whose cellular localization needs to be regulated during signaling. It is possible that depending on the endocytic process, cargo and stimulus, mechanistically different pathways are used. For a further discussion of this idea, please refer to the recent review of Otto and Nichols (71). Another important question is how the modifications (such as phosphorylation) and oligomerization of flotillins affect the trafficking of their putative cargo molecules. This topic is discussed more in detail below in the chapter *"Functional significance of phosphorylation and oligomerization of flotillins"*.

Functions of flotillins in cell migration and adhesion

Already the discovery of flotillins by using antibodies directed against cell surface proteins led to the assumption of a functional role of flotillin proteins in cell adhesion (2, 82). Up to date, much has been speculated about this attribute, but only a few studies actually showed an involvement of flotillins in the establishment and maintenance of cell-cell and cell-matrix adhesion structures. One of the first functional hints was obtained in the study by Hoehne *et al.* (83), in which they investigated the functional role of flotillins in the model organism *Drosophila melanogaster*. The null mutant of flotillin-2 in *Drosophila* is viable and showed no visible phenotype, whereas Mz1369-Gal4 regulated overexpression of flotillin-2 alone or together with flotillin-1 resulted in a severe disturbance of the ommatidial pattern in regard to the number of primary pigment cells and cone cells. Furthermore, staining with antibodies against the cell adhesion molecules Roughest, Kin-of-irre and Sticks-and-Stones of the IgCAM protein family showed a strong labeling of an abnormal accumulation of multivesicular bodies in interommatidial cells. These observations led to the conclusion that flotillins interfere with the distribution of cell adhesion and signaling molecules in imaginal discs of *Drosophila*. In 2009, Málaga-Trillo *et al.* (15) published that the loss of the prion protein (PrP) in zebrafish embryos is characterized by the loss of embryonic cell adhesion. They presented confocal images of drosophila S2 cells, in which overexpressed mouse PrP-EGFP colocalized with overexpressed rat flotillin-2-dsRed at cell-cell contact sites. Similar results were obtained with overexpressed zebrafish PrP-2-EGFP with rat flotillin-1-dsRed and flotillin-2-dsRed as well as zebrafish flotillin-1a-dsRed. (15)

One major type of cell-cell contact in multicellular organisms is the adherens junction. The core of adherens junctions is composed of proteins of the cadherin and catenin family (84). Insights into a functional association of flotillins with this cell-cell contact type were obtained in the study of Bodrikov *et al.* (85), who showed that a transient knockdown of flotillin-2 in primary hippocampal neurons perturbed axon growth and differentiation. This phenotype was rescued in 62% of the neurons by overexpression of the constitutive active mutant of the GTPase TC10 which participates in the exocyst dependent polarized delivery of cargo to the growth cone (86). Interestingly, the amount of coprecipitated of N-Cadherin, which is a core protein of adherens junctions in adult neuronal cells, increased in flotillin-2 immunoprecipitates from mouse hippocampal HPL3-4 cells after application of PrP-Fc. Overexpression of a dominant negative TC10 blocked this effect, which indicates that the PrP together with flotillin-2 is important for the recruitment of N-Cadherin, which in turn takes place in a TC10-dependent manner (85). Very recently, the Stuermer group could show that flotillins together with the PrP play a role in E-Cadherin recruitment to adherens junctions (79). They showed that both proteins colocalized with E-cadherin at the plasma membrane in human epidermoid carcinoma A431 cells. Knockdown of flotillin-2 or PrP resulted in an abnormal shape of adherens junctions, causing overlapping of neighboring cells (79). In addition, electron microscopy revealed that the length of adherens junctions was shorter in flotillin-2 and PrP knockdown cells (79).

The complexing of the junctional proteins p120-catenin (p120ctn) and N-Cadherin at cell-cell contact sites was shown to occur in cholesterol rich membrane domains in mouse C2C12 cells (87). A similar observation was made in differentiated human colon adenocarcinoma HT-29 cells, in which the interaction of E-Cadherin with p120ctn preferentially takes place in lipid rafts. A stable knockdown of flotillin-1 resulted in a diffuse localization of p120ctn and an impaired recruitment of p120ctn and E-Cadherin in lipid rafts. In addition, flotillin-1 depletion decreased the enzymatic activity of ALP and DPPIV during the enterocytic differentiation process (88). This study revealed a novel function of flotillin-1 in a lipid raft mediated maturation of adherens junctions and in intestinal cell differentiation.

Other studies implicated that flotillins seem to be of functional relevance for cell-matrix adhesion processes. We showed that the transient knockdown of flotillin-2 in HeLa cells impaired cell spreading on fibronectin, as compared to cells transfected with control siRNA (30). Overexpression of flotillin-2-EGFP enhanced cell spreading and induced filopodia-like protrusions in an expression-level dependent manner in several epithelial cells lines, e.g. HaCaT cells (4, 30). The observed enrichment of flotillin-2 fusion proteins at cell-cell contact sites and the induction of filopodia suggest an active link to the cytoskeleton. Indeed, use of *in vivo* colocalization assays, *in vitro* binding assays and actin polymerization assays suggested that flotillin-2 interacts with F-actin via its SPFH domain, and that this binding mediates the anchoring of the actin cytoskeleton to the plasma membrane (36).

Role of flotillins in EGF receptor and mitogen activated protein kinase signaling

In the last 12 years, various studies have shown that flotillin-1 and flotillin-2 are physically and functionally linked to signal transduction pathways of several membrane receptors, especially receptor tyrosine kinases such as the insulin receptor (IR), the nerve growth factor (NGF) receptor, tropomyosin-related kinase A (TrkA), the polymeric Immunoglobulin E (IgE) receptor, fibroblast growth factor receptor (FGFR) and the epidermal growth factor receptor (EGFR) (27, 30, 63, 65, 89, 90). However, the mechanisms how flotillins may regulate the signaling process have not been dissected in detail. Only recently, we have shed light on the molecular mechanisms of flotillin-1 function in EGFR/MAPK signaling (27, 89). We showed that flotillin-1 actually plays a dual role by first regulating the clustering of EGFR in the plasma membrane after EGF stimulation and, later on during signaling, by functioning as a MAPK scaffolder protein (27). We also demonstrated that flotillin-1, but not flotillin-2, interacts with Fibroblast Growth Factor Receptor Substrates 2 and 3 (FRS2 and FRS3) (89) which have previously been shown to be involved in MAPK signaling (91-93). (See Figure 3).

Depletion of flotillin-1 in HeLa cells results in a decreased tyrosine phosphorylation of the EGFR especially at Tyr1173 already after a short EGF stimulation, and the reduced phosphorylation persists up to a time point where no phosphorylation of EGFR can be detected (27). By means of TIRF microscopy, we showed that the formation of EGFR clusters at the cell surface upon EGF stimulation is impaired after flotillin-1 knockdown. This fits well with the findings of Schneider *et al.* (75) who showed that flotillins might be important for the clustering of APP on the cell surface before its endocytosis. In addition to EGFR clustering, flotillin-1 depletion also prevents the association of EGFR and some of its downstream signaling partners with membrane rafts (27). Although the fraction of total cellular EGFR that is found in rafts is very low, this might be the population of the receptors that is capable of fast signaling upon stimulation. Earlier findings from Hofman *et al.* have suggested that the raft associated EGFR actually resides in two separate populations of rafts which then coalesce after stimulation, bringing the required signaling molecules in close

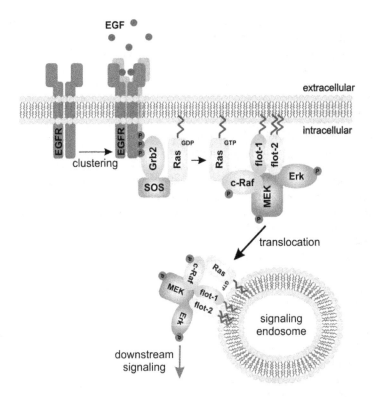

Figure 3. Flotillin-1 is a MAPK scaffolder - the role of flotillins in EGFR/MAPK signaling. Binding of the ligand EGF results in dimerization and activation of the EGFR. Thereby, several Tyr residues in its kinase domain become either autophosphorylated or phosphorylated by other kinases. These phosphorylation events lead to the recruitment of further signaling components, especially those of the MAPK signaling pathway. Grb2 binds to the phosphorylated EGFR kinase domain and transmits the downstream signal via Ras, c-Raf and MEK to the MAPK ERK1/2. The EGF-induced signaling through MAPK is initiated at the plasma membrane but later on translocates to specialized endosomes, i.e. signaling endosomes. Flotillins are able to associate with the EGFR itself and regulate the clustering of the receptor at the plasma membrane. Flotillin-1 directly interacts with the components of the MAPK signaling pathway, e.g. c-Raf, MEK, KSR1 and ERK and can thus be classified as a novel MAPK scaffolder.

contact (94). This model is well in line with our earlier findings showing that upon EGF stimulation, the molecular mass of flotillin oligomers increases, most likely due to their coalescence into larger complexes (56).

We also observed a reduced downstream signaling towards the canonical MAPK pathway and Protein Kinase B/Akt after stimulation with EGF or basic fibroblast growth factor (bFGF) upon flotillin-1 knockdown (27), implicating that flotillin-1 is not only involved in EGFR but also in FGFR signaling. Similar results with EGF have later on been obtained by

Solis *et al.* (79) in A431 cells, which overexpress EGFR to a very high degree, after flotillin-2 depletion. This is somewhat in contrast to our unpublished results in HeLa cells, which express physiological amounts of EGFR. In these cells, depletion of flotillin-2, resulting in nonexpression of flotillin-1, actually causes an increased MAPK phosphorylation upon EGF stimulation (A. Banning & R.T., unpublished findings). This again stresses the importance of using analogous cell systems for studies in signaling, as the degree of downstream signaling is highly dependent on the amount of EGFR expressed.

Flotillin-1 depletion also resulted in reduced activation of the MAPK pathway, especially of ERK kinases, which are required for transmitting the MAPK signal into the nucleus (27). Although flotillin-1 depletion clearly impairs an early step, EGFR clustering, during the signaling, which alone would be sufficient to inhibit MAPK signaling, this appears not to be the only reason for the reduced ERK activity. Incubation of the cells with phorbol myristate acetate (PMA) results in the activation of PKC, which in turn can active the cRAF kinase in the absence of an upstream EGFR signal (95, 96). Our attempt to overcome the downstream ERK inhibition by PKC-mediated activation of cRAF resulted in normal activation of cRAF and MEK kinases, whereas the ERK phosphorylation could not be rescued in flotillin-1 knockdown cells (27). This demonstrates that flotillin-1 must play a role not only at the plasma membrane but also at the level of ERK activation. Accordingly, we showed that flotillin-1 directly interacts with several MAPK pathway proteins, namely cRAF, MEK1 and ERK2. This interaction was not dependent on the MAPK scaffolder Kinase Suppressor of Ras (KSR), since flotillin-1 was found to coprecipitate with the MAPK component also in KSR1 depleted cells (27). Thus, these results suggest that flotillin-1 is indeed a *bona fide* MAPK scaffolder that is capable of directly interacting with several MAPK pathway proteins. Interestingly, the flotillin-1 interactor FRS2 has also been suggested to function as a scaffolding protein for the MAPK proteins (92, 97). Furthermore, our yeast two-hybrid studies have revealed another MAPK scaffolding protein, Mitogen-activated Protein Kinase Organizer 1 (MORG1) as an interaction partner of flotillins (our unpublished data). In future studies, it will be interesting to dissect the details of the regulation of MAPK signaling by flotillin-1, e.g. how these complexes are regulated during signaling in terms of their molecular composition and cellular localization.

The role of flotillins in EGFR signaling is supported by our recent findings in stable flotillin-1 and flotillin-2 knockdown human breast cancer MCF-7 cells (Figure 4). Comparable to HeLa cells, in MCF7 cells the stability of flotillin-1 is strongly dependent on flotillin-2, since the knockdown of flotillin-2 resulted also in a loss of flotillin-1. However, knockdown of flotillin-1 had no effect on flotillin-2 expression. Immunoblot analysis showed that the amount of EGFR is increased in flotillin-1 and flotillin-2 knockdown cells (Figure 4A). This effect is only visible in stable knockdown cells and not in transient knockdown cells, most likely due to an adaption of the EGFR amount to overcome the decreased EGFR-MAPK signaling in flotillin knockdown cells. We have previously observed this effect also in HeLa cells and other cell lines (our unpublished results). The EGFR, also known as ErbB1/Her1, is one of the four known members of the ErbB/Her protein-tyrosine kinase family (reviewed in (98)). Analysis of the protein amount of ErbB2/Her2 and ErbB3/Her3 showed no detectable

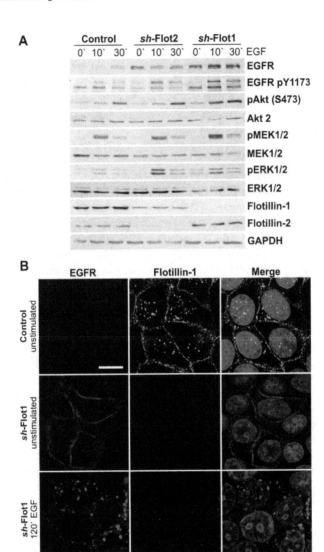

Figure 4. The reduced activation of the MAPK signaling pathway due to fiotillin depletion is
compensated by an increased protein expression level of the epidermal growth factor receptor (EGFR)
in MCF-7 cells. (A) Stimulation of stable flotillin-1 and flotillin-2 MCF-7 knockdown cells with EGF for
10 minutes and 30 minutes. The reduced activation of the MAPK signaling in flotillin knockdown cells
is compensated by an upregulation of the EGFR protein level. (B) Immunofluorescence of endogenous
EGFR (red) and endogenous flotillin-1 (green) in unstimulated and EGF-stimulated (120 min) flotillin-1
knockdown cells (sh-Flot1). Control cells are shown only in unstimulated state, due to undetectable
EGFR level by immunofluorescence. Scale bar: 20 μm

change in the protein level (data not shown), which indicates that flotillin depletion specifically affects the protein level of the EGFR. In general, the endogenous EGFR is not detectable by immunofluorescence in MCF-7 cells, whereas in flotillin-1 depleted MCF-7 cells, a prominent staining at the plasma membrane is visible (Figure 4B). Stimulation with EGF for 120 minutes results in a translocation of the EGFR in vesicular structures in these cells. This is in line with our previous results in HeLa cells, indicating that the endocytosis of the EGFR is not affected by flotillin-1 depletion (27).

Flotillins and other receptor tyrosine kinases

The first experimental evidence for a functional role of flotillins in insulin signaling was presented by the Saltiel group (63). They demonstrated that flotillin-1 forms a ternary complex with c-Cbl-associated protein (CAP) and the E3 ubiquitin protein ligase Cbl upon insulin stimulation (Figure 5). This was shown to be important for lipid raft association, downstream signaling of the CAP-Cbl complex and finally for glucose uptake by the glucose transporter GLUT4 in 3T3-L1 adipocytes (63). Previously, the interaction of CAP with Cbl was shown to be independent of insulin stimulation *in vivo* and *in vitro* and to be mediated by SH3 domains (99, 100). In line with this, expression of CAPΔSH3 inhibited the translocation of GLUT4 from storage vesicle to the plasma membrane upon insulin stimulation (63). Further analysis of the flotillin-1 interaction mode with murine CAP isoforms showed that flotillin-1 only interacts with the isoform CAP-4 via its first hydrophobic domain, but not with the isoforms CAP-1, -2, and -3 (31). In CAP, the sorbin homology domain (SoHo) seems to be important for flotillin-1 binding, as the expression of CAPΔSoHo prevented the translocation of Cbl to lipid rafts in 3T3-L1 adipocytes (101). Since the murine CAP isoforms 1-4 all contain the SoHo domain, a predicted 70 amino acids long proline-rich region N-terminal to the SoHo domain in CAP-4 might also be of importance for flotillin-1 binding (102). While flotillin-1 was found in perinuclear regions in differentiated skeletal muscle cells (103), it was located at the plasma membrane in adipocytes (31) under basal conditions, suggesting that the role of flotillin-1 in targeting GLUT4 to the plasma membrane after insulin stimulation may differ between muscle and adipose tissue. Indeed, Fecchi *et al.* (103) showed that flotillin-1 colocalized with GLUT4 in the perinuclear region of myotubes, and both proteins translocated to the sarcolemma in a PI3K/ PKCζ- and caveolin-3 dependent manner upon stimulation with insulin. In addition, it was demonstrated that caveolin-3 is involved in the PI3K-dependent insulin signaling pathway, whereas flotillin-1 is crucial for the PI3K-independent insulin signaling pathway by interacting with the main scaffolding protein CAP. The interaction with CAP was also shown for flotillin-1 and flotillin-2 in mouse neuroblastoma N2a cells, in which overexpressed flotillin-1 and flotillin-2 colocalized with endogenous CAP at focal adhesion sites (57). This result indicates that by means of their interaction with CAP, flotillins might play an additional role in focal adhesion dynamics and integrin signaling.

Receptor tyrosine kinases of the Trk receptor family regulate various cellular processes, such as proliferation and differentiation, through several cellular signal cascades. Three types of Trk receptors exist in humans, including TrkA, TrkB and TrkC, which have varying affinities to neutrotrophins such as NGF. TrkA is specifically activated by NGF (reviewed in (104, 105). Downstream signaling by TrkA seems to be dependent on the lipid raft association of TrkA. Upon NGF stimulation, TrkA is recruited to lipid rafts in PC12 cells, a process dependent on its association with CAP, which in turn binds to flotillin-1 (65). The deletion of the CAP SoHo domain abolishes the lipid raft association of TrkA and further downstream signaling events such as ERK1/2 phosphorylation (65). (See Figure 5)

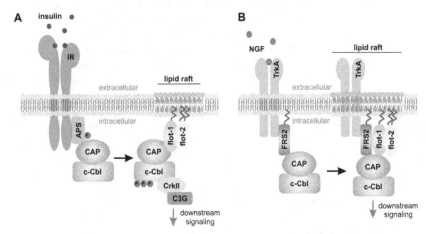

Figure 5. Proposed functional role of flotillin proteins in insulin and NGF signaling cascades. (A) Insulin is the major hormone controlling essential metabolic processes such as glucose and lipid turnover. Insulin binding to the extracellular subunits of the insulin receptor tyrosine kinase (IR) induces a conformational change and results in the autophosphorylation of several tyrosine residues in the cytoplasmic part of the IR. In adipocytes, the constitutive complex of the E3 ubiquitin ligase c-Cbl with the c-Cbl associated protein (CAP) is recruited to the active receptor. The interaction of CAP/c-Cbl complex with the IR was found to be indirect and mediated via the Adapter protein with Pleckstrin homology and Src homology 2 domains (APS). To initiate downstream signaling events, the CAP/c-Cbl complex then dissociates from the IR and is recruited to lipid rafts by forming a ternary complex with flotillin-1. (B) Receptor tyrosine kinases of the Trk receptor family regulate various cellular processes in the mammalian neuronal system, such as proliferation and differentiation by activation through neutrophins. The receptor TrkA is activated by the neutrophin nerve growth factor (NGF), which leads to the recruitment of the CAP/c-Cbl complex to TrkA. Since the fibroblast growth factor substrate 2 (FRS2) is also capable of binding to TrkA and CAP, it might be possible that the CAP/c-Cbl complex is recruited to TrkA by its interaction with FRS2. Most likely by the scaffolding activity of flotillin-1, the complex including the TrkA receptor is translocated to lipid rafts for a proper initiation and amplification of the downstream signal.

Recently, we have demonstrated that flotillin-1 directly interacts *in vitro* and *in vivo* with FRS2, a protein suggested to function in insulin signaling (106), by means of its N-terminal phosphotyrosine binding (PTB) domain and competes for the binding with FGFR (89). Interestingly, FRS2 was shown to interact directly through its PTB domain with TrkA in a phosphotyrosine-dependent manner after NGF stimulation (107, 108). Furthermore, we showed that in addition to flotillin-1, FRS2 also binds CAP by means of both its PTB domain and the C-terminal domain. In CAP, the SoHo domain in cooperation with the third SH3 domain mediate FRS2 binding (89). Both flotillin-1 and CAP bind to the PTB domain of FRS2, which suggests that CAP and flotillin-1 may compete for FRS2 binding to regulate the formation of signaling complexes.

Animal models for the function of flotillins

Flotillins are highly conserved proteins which are assumed, mainly on the basis of cell culture studies, to play multifaceted roles in developmental regulation. Hoehne *et al.* carried out initial studies of loss- and gain-of function of flotillins in wild type and mutant *Drosophila melanogaster* to investigate their role during morphogenesis and organogenesis (83). According to this study, flotillin proteins are highly expressed in embryonic neuronal tissues, and during later developmental stages flotillins showed an enhanced expression pattern in axon fascicles (83). Additionally, flotillin specific antibodies labeled in particular the *Drosophila* mushroom body, which is a paired structure of insect central brain responsible for higher-order sensory integration and learning. Flotillins showed a categorical neuronal expression, as antibody staining was restricted to optic lobes and central brain but was not observed in glial cells. A P-element-induced null mutant for flotillin-2 was viable, fertile and with no apparent deficits in behavior or overall activity. In this flotillin-2 null mutant, the expression of flotillin-1 was undetectable even though the mRNA transcript was present (83). Conversely, overexpression of flotillin-1 and flotillin-2 proteins either alone or together in the eye imaginal discs lead to mislocalization of adhesion molecules of the immunoglobulin superfamily (IgCAMs; please also refer to the chapter *"Functions of flotillins in cell migration and adhesion"*). Additionally, the adult flies showed a variable phenotype affecting bristles, ocelli, eyes, wings, abdominal segments and thoracic structures. A severe wing phenotype, including blistering and melanotic cells, was caused by flotillin overexpression in wing imaginal discs. Moreover, a later study showed that flotillin-2 deficiency resulted in reduced spreading of the morphogens Wnt and hedgehog, whereas flotillin-2 overexpression resulted in increased morphogen secretion and expanded diffusion. These findings qualify flotillin-2 as a novel component of the Wnt and hedgehog secretion pathway in *Drosophila* (64).

Flotillins are also implicated to regulate axon regeneration in a zebrafish model. As a consequence of gene duplication in fish, there are two copies of each flotillin gene. To understand the essential role of flotillins in axon regeneration, specific morpholino (Mo) antisense nucleotides were used to transiently downregulate flotillin-1a, flotillin-1b and flotillin-2a in zebrafish (60). Interestingly, the number of retinal ganglion cells (RGC) undergoing axon regeneration was reduced by 69% in flotillin-Mo treated eyes after 7 days (60). Recently, a zebrafish model was employed to study the role of flotillins in cholera intoxication. Consequently, flotillin-1 and flotillin-2 knockdown prevented cholera intoxication in zebrafish embryos. Moreover, cholera toxin subunit B (CTxB) colocalized with flotillins at the plasma membrane and in endosomes in COS1 cells (109). Thus, these results suggest flotillins as essential elements for axon regeneration and indicate a novel cholera intoxication *in vivo* model dependent on flotillins for trafficking between plasma membrane/endosomes to the ER in the zebrafish.

Cell migration is an essential mechanism during the development and maintenance of multicellular organisms. It requires the constant turnover of cell-matrix adhesion structures, e.g., integrin-based focal contacts. Results presented by Ludwig *et al.* (110) strongly suggest a functional role of flotillin microdomains during neutrophil migration, in uropod formation, and in the regulation of myosin IIa activity. They generated and characterized C57BL6/J mice lacking exon 3-8 of the gene for flotillin-1. In line with the observations in *Drosophila*, these flotillin-1$^{-/-}$ mice are viable, fertile, and have no readily apparent phenotype. On the cellular level, absence of flotillin-1 resulted in a greatly reduced flotillin-2 expression, and the residual flotillin-2 was not found in detergent-resistant fractions of flot-1 $^{-/-}$ MEF cells. These data suggest that deletion of flotillin-1 leads to the abrogation of flotillin microdomain function. Furthermore, isolated flotillin-1$^{-/-}$ neutrophils exhibited a defect in chemotaxis when exposed to the chemoattractant N-formyl-Met-Leu-Phe (fMLP). Previous reports have already shown that, upon stimulation with chemoattractants, flotillin microdomains rapidly distribute to the uropod, a contractile structure at the back of leukocytes (32, 111). This suggested that these microdomains could participate in polarization and/or chemotaxis. Ludwig *et al.* were also able to show by mass spectrometry of isolated flotillin microdomains and immunoprecipitation of flotillin-2-EGFP an association of flotillin microdomains with myosin IIa, α-spectrin and β-spectrin. Flotillin overexpression led to an increase of myosin IIa activity, whereas the loss of flotillins caused a reduction as measured by using phosphospecific antibodies (110). The regulatory effect of flotillins on myosin IIa activity strengthens the assumption that both proteins exhibit a functional role during cell migration, as myosin IIa was shown to act as a negative regulator of epithelial cell migration, and its ablation results in severe defects in focal adhesion formation and actin stress fiber organization (112, 113).

4. Functional significance of phosphorylation and oligomerization of flotillins

Recent data from several groups have revealed flotillins as important regulators of vital cellular processes such as signaling and endocytosis. One major challenge in the future will be to dissect how the oligomerization and the covalent modifications of flotillins affect their function during these processes. Endocytosis of flotillins appears to be influenced by upstream signaling (30, 49, 56) and requires clustering of flotillin-1 and flotillin-2 molecules (34, 56). We have shown that EGF stimulation results in uptake of flotillins hetero-oligomers from the plasma membrane, and that hetero-oligomerization appears to be required for the endocytosis (30, 56). Mutation of a single Tyr residue, Tyr163, in flotillin-2 is sufficient to prevent the EGF-induced endocytosis of flotillins (30), and this mutation impairs the hetero-oligomerization of flotillin-2 with flotillin-1 (56). Flotillins are phosphorylated by the Src family kinases cSrc and Fyn (4, 49), but the direct role of Tyr phosphorylation in the endocytosis is still under debate. Our data show that inhibition of cSrc kinase, the activity of which is necessary for flotillin-2 phosphorylation (30), does not impair the endocytosis of flotillin-2 (16), whereas the mutation of the single Tyr residue Y163 does (30). Thus, it is more likely that this mutation affects the oligomerization of flotillins and thereby also endocytosis. However, phosphorylation of flotillins by the Fyn kinase appears to be sufficient to induce endocytosis of flotillins (49), suggesting that the Src family kinases might play different roles in terms of flotillin function. It is also important to keep in mind that PKC has recently been shown to phosphorylate flotillin-1 in Ser315, and that this phosphorylation is required for the endocytosis of the dopamin transporter DAT (74). Thus, phosphorylation seems to be an important direct or indirect regulator of flotillin trafficking.

Not only phosphorylation but also covalent modifications with fatty acids myristate and palmitate take place in the case of flotillins (3, 4, 53). Mutation of Gly2 prevents the myristoylation and the subsequent palmitoylation of flotillin-2, thus also inhibiting its membrane association and rendering the protein soluble (4). Recently, the palmitoyl transferase DHHC5 was shown to palmitoylate flotillin-2 (53). Interestingly, growth factor withdrawal, which induces neuronal differentiation, resulted in rapid degradation of DHHC5 (53), implicating that palmitoylation might be important for the well established function of flotillins in neurite outgrowth in cultured neuronal cells. Since the soluble, non-modified flotillin-2 with Gly2Ala mutation seems to function as a dominant negative protein during cell-matrix adhesion (30), and successive removal of the three palmitoylation sites affect the membrane association of flotillin-2 (4), the fatty acid modifications, especially the reversible palmitoylation, are most likely important for the regulation of flotillin function. Accordingly, we have observed that flotillin-2 expression is necessary for the membrane association of flotillin-1, which is not myristoylated and only palmitoylated in a single residue. Thus, ectopic expression of flotillin-1 in flotillin-2 depleted cells results in a soluble protein that is not capable of exerting its function (our unpublished data).

Our recent findings revealed flotillin-1 as a MAPK regulator (27). However, no data are so far available on the role of flotillin modifications such as phosphorylation in the regulation of MAPK signaling. Intriguingly, another SPFH/PHB family protein, prohibitin-1, has been shown to regulate an earlier step during MAPK signaling, namely signaling by the small GTPase Ras and RAF kinase (114, 115). Although the most evident cellular localization of prohibitin is the inner mitochondrial membrane, various studies have now shown that prohibitin is a substrate of several cytoplasmic kinases such as Akt, and it becomes Tyr phosphorylated upon insulin stimulation of cells (116, 117). Thus, at least a fraction of prohibitin must be found in another cellular localization such as the plasma membrane. Importantly, prohibitin has been shown to be essential for cell migration that is induced by the Ras-Raf-MAPK pathway (115), and very recent findings show that phosphorylation of prohibitin, which takes place in a raft domain, is necessary for the Ras-dependent enhancement of cellular metastasis (118), verifying the results of Rajalingam *et al.* (115). Thus, there are several parallels in prohibitin and flotillin function, including MAPK signaling and cell migration, although they appear to be involved in different steps of the signaling. Since some evidence for a hetero-oligomerization of different SPFH-PHB family members, namely prohibitin and stomatin family protein SLP-2 (119) has been presented, it would be important to study if flotillins and prohibitins are capable of hetero-oligomerizing with each other and how this affects their function, especially with respect to the MAPK signaling. In the future, it will also be of interest to study if the phosphorylation-incompatible forms of flotillins cause aberrant signaling and if this effect is linked on phosphorylation at specific sites. Our previous data show that the function of flotillin-2 in cell-matrix association is indeed dependent on specific Tyr residues, the mutations of which affect cell spreading (30). Similarly, mutations of phosphorylated residues in flotillin-1 might affect its role in MAPK signaling and its interaction with the MAPK proteins.

5. Flotillins in human diseases

Role of flotillins in cancers

Early findings already indicated that flotillins may be involved in the regulation of the actin cytoskeleton. Overexpression of flotillin-2 was found to result in induction of filopodia-like protrusions and changes in the cytoskeleton (4, 11). Furthermore, flotillins have been shown to be important for the polarization of various immune cells such as T cells and neutrophils (29, 32, 33, 37, 110, 111), and to be involved in the regulation of the activity of various small GTPases that control e.g. the remodeling of the actin cytoskeleton (36, 57). In fact, flotillin-2 has even been suggested to directly interact with actin (36), although this has not been verified in further studies. Thus, there are very strong implications that flotillins are required for a proper regulation of the actin cytoskeleton, which is necessary for e.g. the control of cell migration. Consequently, taking also into account the numerous reports on the involvement of flotillins in signaling, it is evident that flotillins are critical factors that regulate the cell physiology and cell growth. Thus, the findings showing that flotillins are involved in some types of cancer are well in accordance with this function.

The strongest evidence for the role of flotillins in cancer to date comes from studies on human malignant melanoma. The direct association of flotillins with human cancers was shown in 2004 when the group of M. Duvic demonstrated that flotillin-2 is highly expressed in various melanoma cell lines, and that flotillin-2 expression correlated with the progression of human melanoma (120). Furthermore, the thickness of primary melanoma lesions, so-called Breslow depth, which is a prognostic marker for the malignancy of a melanoma lesion, showed a correlation with the increased flotillin-2 expression (121). Importantly, benign, non-malignant cells could be converted into tumorigenic, metastatic ones upon overexpression of flotillin-2 (120). Furthermore, flotillin-2 was shown to be associated with the thrombin receptor PAR1, the expression of which was dependent on the level of flotillin-2 protein (120).

Lin *et al.* detected a high expression of flotillin-1 in the majority of the breast cancer specimens they studied (66). Furthermore, they revealed a significant correlation between the expression level of flotillin-1 and poor survival of the patients, suggesting that flotillins might prove useful as novel diagnostic markers for cancer. Knockdown of flotillin-1 was shown to result in inhibition of Akt activity and increased transcriptional activity of the transcription factor FOXO3a. Knockdown of flotillin-1 also inhibited the proliferation of breast cancer cell lines by inducing a G1-S phase arrest of the cell cycle. In line with this, the expression of the cyclin dependent kinase (CDK) regulator cyclin D1 was reduced after flotillin-1 silencing, whereas that of the CDK inhibitors p21[Cip1] and p27[Kip1] was considerably increased (66), suggesting that flotillin-1 is an important regulator of the cell cycle. This result is strongly supported by our findings showing that in flotillin-1 knockdown cells, cyclin D1 expression after EGF stimulation is severely impaired (27).

Although flotillins evidently play a role in cancer and their overexpression is associated with malignancy, very little is known about their transcriptional regulation. Flotillin-2 was shown to be a transcriptional target of the p53 family transcription factors p63 and p73 but not of p53 (122). We have recently addressed this question more in detail and shown that flotillins are transcriptionally regulated by the Extracellularly Regulated Kinase signaling and by the Retinoid X Receptor (Banning *et al.*, in revision). Due to their important role in cancer, a deeper knowledge on the transcriptional regulation of flotillins is essential for understanding the mechanisms of malignancy that results from flotillin upregulation.

Implications of flotillins in Alzheimer's Disease and other neurodegenerative disorders

Alzheimer's disease (AD) is the most common age-related neurodegenerative disorder whose most prominent pathological hallmarks include neuronal loss, neurofibrillary tangle formation and senile plaques throughout the brain cortex. The major component of the senile plaques is the amyloid β peptide (Aβ) comprising of 40-42 residues, which is generated by a proteolytic cleavage of a large transmembrane precursor protein named amyloid β precursor protein (APP). APP is subjected to proteolytic processing at several sites by a group of proteases called secretases. Sequential cleavage of APP by β-secretase (β–site APP cleaving enzyme or BACE) and γ-secretases generates Aβ and is hence termed the amyloidogenic pathway (reviewed in (123-126)). APP can also be processed through a non-amyloidogenic pathway, in which the α-secretase cleaves within the Aβ sequence, thereby precluding the formation of Aβ. It has been strongly implicated that the amyloidogenic processing of APP takes place in lipid rafts, which is supported by the findings showing that APP can be detected in rafts isolated from cortical neurons (124). Besides APP, other AD associated raft resident proteins include BACE-1 (127), endoproteolytic fragments of presenilin-1 (PS-1) and the presenilins (128, 129). Compelling evidence from various research groups shows that the flotillin family of proteins is partially associated with AD pathology and trafficking (47, 130, 131). Flotillins are considered as classic raft markers in neuronal tissue since they show an abundant expression in the brain, typically in pyramidal neurons and astrocytes. Kokubo *et al.* detected flotillin-1-positive rafts in the plasma membrane and membranes of intracellular organelles in rat brain tissue by using immunoelectron microscopy (47). Consequently, they showed that in APP transgenic mice, flotillin-1 and Aβ42 colocalized in approximately 10% of flotillin-1 positive rafts, and that flotillin-1 expression increased with the accumulation of Aβ (132).

Parallel studies in brain sections from AD, Down syndrome and non-demented subjects with plaques showed enhanced flotillin-1 expression with the progression of AD (130). Further insight was provided by Rajendran *et al.* who demonstrated that in AD patients, Aβ accumulates and is located in flotillin-1 positive endosomes. They also showed that flotillin-1 was not only linked with intracellular Aβ in transgenic mice but was also present in extracellular plaques of AD patient brain sections (131). Furthermore, comparison of flotillin-1 immunoreactivity in the hippocampus, the amygdala and the isocortex of AD and control patients implicated that flotillin-1 principally accumulates in lysosomes of tangle bearing neurons during the advanced stages of AD (133). To date, the precise functional significance of APP and flotillin association is enigmatic. Schneider *et al.* proposed that flotillins could function as scaffolding proteins, forming a platform for clustering of APP and consequent endocytosis. Accordingly, flotillin-2 knockdown in mouse neuroblastoma cells impaired APP endocytosis and processing, resulting in reduced $A\beta$ production (75). It has previously been shown that the intracellular domain of APP directly interacts with flotillin-1 (134). Our unpublished findings suggest that both flotillin-1 and flotillin-2 directly interact with APP, suggesting that both flotillins could facilitate recruitment of APP into microdomains.

Flotillins and other lipid raft-associated proteins might also be involved in other human neuronal diseases. The study by Shin *et al.* (135) showed an increased level of the lipid raft associated proteins caveolin-1, -2, and -3 in the spinal cord during the early stage of experimental autoimmune encephalomyelitis (EAE). EAE is an experimental animal model, mostly used in rodents, of inflammatory demyelinating diseases of the central nervous system (CNS), including multiple sclerosis (reviewed in (136)). Similar results with the caveolin data were obtained for flotillin-1 during the peak stage of EAE. Immunohistochemical analysis showed an enhanced expression of flotillin-1 in the dorsal horn lamina and in some flotillin-1-positive macrophages and astrocytes, in which it colocalized with the lysosomal marker cathepsin D (137). Due to that, they postulated that flotillin-1 inherits an important function in immune and neuronal cells within the CNS during EAE, possibly via the activation of signal transduction and lysosomal activity.

Parkinson's disease (PD) is one of the most common, progressive neurological disorders. The pathological hallmark of PD is the extensive loss of dopamine secreting cells within the substantia nigra, particularly affecting the ventral component of the pars compacta (reviewed in (138)), which can lead to uncontrolled muscle contraction and movement, dementia, depression and anxiety. Flotillin-1 staining in rat brain sections showed a prominent labeling in the cytoplasm of catecholamine releasing cells (dopamine, norepinephrine, and epinephrine) and an upregulated gene expression of flotillin-1 in the substantia nigra of Parkinson's brain by RT-PCR (139). Although the molecular function of flotillin-1 in Parkinson's disease is still not known, this study implies that flotillin-1 may be involved in the neuronal changes occurring during Parkinson's disease.

Flotillins and pathogens

During evolution, many pathogens have developed mechanisms to use host-cell lipid rafts as signaling and entry platforms to escape the host immune system (reviewed in (140, 141)). Several studies demonstrated that lipid rafts are often used as entry sites for bacteria, e.g., *Salmonella enterica* (142) or *Listeria monocytogenes* (143). In regard to flotillins, Li *et al.* (144) showed that flotillin-1 accumulates at bacteria entry sites and around intracellular enteropathogenic *Escherichia coli* (EPEC). This bacterium causes diarrheal disease by disrupting the integrity of tight junctions, which are essential structures for physiological homeostasis and defense against pathogen invasion. Furthermore, they detected a shift of flotillin-1 and occludin, a major protein of tight junctions, from Triton X-100 insoluble fractions to Triton X-100 soluble fractions following EPEC infection. This led to the conclusion that the loss of tight junction barrier function might be accounted to the redistribution of occludin and flotillin-1 after EPEC infection.

Not only bacteria use lipid raft domains as preferred entry sites in host-cells, but also various enveloped and non-enveloped viruses are dependent on an intact lipid raft structure and the presence of cholesterol for successful virus entry (reviewed in (145)). In the case of the retrovirus human immunodeficiency virus-1 (HIV-1), flotillin-1 was suggested to play a role of in the cellular response to the viral infection since the protein level of flotillin-1 was increased in peripheral blood mononuclear cells after treatment with the HIV-1 component gp120 (146). The primary HIV-1 receptor CD4 and the co-receptor CCR5, both of which are important for HIV-1 host-cell entry, were also shown to associate with flotillin-1 containing lipid rafts in monocytes (147). Recently, flotillin-1 was also identified as an interaction partner of the overexpressed rhesus monkey TRIM5α (148). TRIM5α is a member of tripartite motif (TRIM) protein family, and plays an important role during host cell defense against retroviruses such as HIV-1 especially in old world monkeys, but not in humans (reviewed in (149)). The function of flotillin-1/TRIM5α interaction is still unknown, as flotillin-1 knockdown or overexpression did not affect HIV-1 transduction efficiency, leading to the suggestion that flotillin-1 at least plays no role in the post-entry restriction of HIV-1 in fetal rhesus monkey kidney (FRhK4) cells.

Flotillins and diabetes mellitus

Flotillins seem to be of functional relevance in insulin signaling (see above) and the metabolic disease diabetes mellitus, as first demonstrated by Baumann *et al.* (63). A year later, James *et al.* showed an increased protein level of flotillin and caveolin-1 in muscle tissue and myoblast lysates derived from stroke-prone spontaneously hypertensive (SHRSP) rats as compared to male Wistar-Kyoto control rats (150). It is important to note that SHRSP rats are a model of human insulin resistance, defined by the decreased ability of cells or tissues to respond to physiological levels of insulin especially in skeletal muscles and adipose tissue. Insulin resistance is of major importance in several human diseases, e.g. diabetes type 2 (151). The results of both the above studies lead to the assumption that flotillins might play a role in insulin signaling in the skeletal muscle and adipose tissue. Furthermore, during steady state conditions, flotillin-1 was found in perinuclear regions in differentiated skeletal muscle cells, but at the plasma membrane in adipocytes (31). This observation suggested that the role of flotillin-1 in targeting GLUT4 to the plasma membrane after insulin stimulation may be different between muscle and adipose tissue. In fact in skeletal muscle cells, the insulin-stimulated signaling cascade leading to the translocation of GLUT4 to the plasma membrane is dependent on flotillin-1 and the muscle-specific protein caveolin-3 (103).

Lipid rafts also play a role in diabetic xerostomia, or dry mouth. This disease, when associated with diabetes mellitus, is caused by degenerative changes in the salivary glands leading to increased infectious conditions in the oral cavity and in the long-range aggravates the risk of atherosclerosis and cardiovascular diseases (reviewed in (152)). Wang *et al.* (153) tried to clarify the mechanisms leading to diabetic xerostomia by testing the subcellular localization of aquaporin 5 (AQP5), a water channel protein, in parotid glands of steptozotocin-induced diabetic rats during stimulation with the muscarinic agonist cevimeline. In parotid acinar cells derived from control rats, AQP5 colocalized with flotillin-2 in the cytoplasm and at the apical plasma membrane ten minutes after cemivelime treatment. In contrast, in the parotid acinar cells derived from diabetic rats, the translocation of these proteins did not occur. In summary, the results showed that AQP5 translocation is a lipid raft-dependent process and inhibition of the muscarinic agonist-induced translocation might be the cause of diabetic xerostomia.

6. Conclusions

It has become evident that flotillins play a key role in the regulation of cellular signaling and membrane trafficking. Although flotillins are not essential for life in metazoans, including mammals, as shown by the recent studies in flotillin knockout mice, they appear to be extremely important for many processes in cultured cells. This would suggest that especially mammals are capable compensating for the loss of flotillins by upregulating the expression of other proteins and thus overcoming the defects in e.g. signaling and cell adhesion. This feature is observed in both our flotillin-2 knockout mice and cultured cells with a constitutive flotillin knockdown (our unpublished data). In the future, it will be important to study what the molecular mechanisms of the compensation of loss of flotillin function might be. It will also be of major interest to study if the original function of flotillins in neuronal regeneration can also be observed in the mammalian knockout model. Furthermore, the functional significance of flotillin modifications can now be studied in the context of mammalian organisms when knockout/knock-in mice with specific mutations become available. Thus, the recent development of flotillin-1 and flotillin-2 knockout mice has brought the studies of flotillin functions to a new exciting era, and important data on the physiological function of flotillins can be expected in the near future.

Author details

Nina Kurrle, Bincy John, Melanie Meister and Ritva Tikkanen[*]
Institute of Biochemistry, Medical Faculty, University of Giessen, Giessen, Germany

Acknowledgement

The research in the Tikkanen lab is supported by the German Research council DFG (Grants Ti291/6-1 and Ti291/6-2), the Von Behring-Röntgen Foundation and the state of Hessen (LOEWE program "Non-neuronal Cholinergic Systems"), whose support is gratefully acknowledged.

[*] Corresponding Author

7. References

[1] Bickel PE, Scherer PE, Schnitzer JE, Oh P, Lisanti MP, Lodish HF. Flotillin and epidermal surface antigen define a new family of caveolae-associated integral membrane proteins. J Biol Chem. 1997 May 23;272(21):13793-802.

[2] Schulte T, Paschke KA, Laessing U, Lottspeich F, Stuermer CA. Reggie-1 and reggie-2, two cell surface proteins expressed by retinal ganglion cells during axon regeneration. Development. 1997 Jan;124(2):577-87.

[3] Morrow IC, Rea S, Martin S, Prior IA, Prohaska R, Hancock JF, et al. Flotillin-1/reggie-2 traffics to surface raft domains via a novel golgi-independent pathway. Identification of a novel membrane targeting domain and a role for palmitoylation. J Biol Chem. 2002 Dec 13;277(50):48834-41.

[4] Neumann-Giesen C, Falkenbach B, Beicht P, Claasen S, Luers G, Stuermer CA, et al. Membrane and raft association of reggie-1/flotillin-2: role of myristoylation, palmitoylation and oligomerization and induction of filopodia by overexpression. Biochem J. 2004 Mar 1;378(Pt 2):509-18.

[5] Salzer U, Prohaska R. Stomatin, flotillin-1, and flotillin-2 are major integral proteins of erythrocyte lipid rafts. Blood. 2001 Feb 15;97(4):1141-3.

[6] Simons K, Ikonen E. Functional rafts in cell membranes. Nature. 1997 Jun 5;387(6633):569-72.

[7] Simons K, Gerl MJ. Revitalizing membrane rafts: new tools and insights. Nat Rev Mol Cell Biol. 2010 Oct;11(10):688-99.

[8] Melkonian KA, Ostermeyer AG, Chen JZ, Roth MG, Brown DA. Role of lipid modifications in targeting proteins to detergent-resistant membrane rafts. Many raft proteins are acylated, while few are prenylated. J Biol Chem. 1999 Feb 5;274(6):3910-7.

[9] Fernow I, Icking A, Tikkanen R. Reggie-1 and reggie-2 localize in non-caveolar rafts in epithelial cells: cellular localization is not dependent on the expression of caveolin proteins. Eur J Cell Biol. 2007 Jun;86(6):345-52.

[10] Schroeder WT, Siciliano MJ, Stewart-Galetka SL, Duvic M. The human gene for an epidermal surface antigen (M17S1) is located at 17q11-12. Genomics. 1991 Oct;11(2):481-2.

[11] Hazarika P, Dham N, Patel P, Cho M, Weidner D, Goldsmith L, et al. Flotillin 2 is distinct from epidermal surface antigen (ESA) and is associated with filopodia formation. J Cell Biochem. 1999 Oct 1;75(1):147-59.

[12] Deininger SO, Rajendran L, Lottspeich F, Przybylski M, Illges H, Stuermer CA, et al. Identification of teleost Thy-1 and association with the microdomain/lipid raft reggie proteins in regenerating CNS axons. Mol Cell Neurosci. 2003 Apr;22(4):544-54.

[13] Glebov OO, Bright NA, Nichols BJ. Flotillin-1 defines a clathrin-independent endocytic pathway in mammalian cells. Nat Cell Biol. 2006 Jan;8(1):46-54.

[14] Stuermer CA, Lang DM, Kirsch F, Wiechers M, Deininger SO, Plattner H. Glycosylphosphatidyl inositol-anchored proteins and fyn kinase assemble in noncaveolar plasma membrane microdomains defined by reggie-1 and -2. Mol Biol Cell. 2001 Oct;12(10):3031-45.

[15] Malaga-Trillo E, Solis GP, Schrock Y, Geiss C, Luncz L, Thomanetz V, et al. Regulation of embryonic cell adhesion by the prion protein. PLoS Biol. 2009 Mar 10;7(3):e55.

[16] Babuke T, Tikkanen R. Dissecting the molecular function of reggie/flotillin proteins. Eur J Cell Biol. 2007 Sep;86(9):525-32.

[17] Lopez D, Kolter R. Functional microdomains in bacterial membranes. Genes Dev. 2010 Sep 1;24(17):1893-902.

[18] Haney CH, Long SR. Plant flotillins are required for infection by nitrogen-fixing bacteria. Proc Natl Acad Sci U S A. 2010 Jan 5;107(1):478-83.

[19] Otto GP, Nichols BJ. The roles of flotillin microdomains--endocytosis and beyond. J Cell Sci. 2012 Dec 1;124(Pt 23):3933-40.

[20] Edgar AJ, Polak JM. Flotillin-1: gene structure: cDNA cloning from human lung and the identification of alternative polyadenylation signals. Int J Biochem Cell Biol. 2001 Jan;33(1):53-64.

[21] Browman DT, Hoegg MB, Robbins SM. The SPFH domain-containing proteins: more than lipid raft markers. Trends Cell Biol. 2007 Aug;17(8):394-402.

[22] Tavernarakis N, Driscoll M, Kyrpides NC. The SPFH domain: implicated in regulating targeted protein turnover in stomatins and other membrane-associated proteins. Trends Biochem Sci. 1999 Nov;24(11):425-7.

[23] Solis GP, Hoegg M, Munderloh C, Schrock Y, Malaga-Trillo E, Rivera-Milla E, et al. Reggie/flotillin proteins are organized into stable tetramers in membrane microdomains. Biochem J. 2007 Apr 15;403(2):313-22.

[24] Volonte D, Galbiati F, Li S, Nishiyama K, Okamoto T, Lisanti MP. Flotillins/cavatellins are differentially expressed in cells and tissues and form a hetero-oligomeric complex with caveolins in vivo. Characterization and epitope-mapping of a novel flotillin-1 monoclonal antibody probe. J Biol Chem. 1999 Apr 30;274(18):12702-9.

[25] Yokoyama H, Fujii S, Matsui I. Crystal structure of a core domain of stomatin from Pyrococcus horikoshii Illustrates a novel trimeric and coiled-coil fold. J Mol Biol. 2008 Feb 22;376(3):868-78.

[26] Ha H, Kwak HB, Lee SK, Na DS, Rudd CE, Lee ZH, et al. Membrane rafts play a crucial role in receptor activator of nuclear factor kappaB signaling and osteoclast function. J Biol Chem. 2003 May 16;278(20):18573-80.

[27] Amaddii M, Meister M, Banning A, Tomasovic A, Mooz J, Rajalingam K, et al. Flotillin-1/reggie-2 plays a dual role in the activation of receptor tyrosine kinase/map kinase signaling. J Biol Chem. 2012 Jan 9.

[28] Langhorst MF, Reuter A, Jaeger FA, Wippich FM, Luxenhofer G, Plattner H, et al. Trafficking of the microdomain scaffolding protein reggie-1/flotillin-2. Eur J Cell Biol. 2008 Jan 29.

[29] Langhorst MF, Reuter A, Luxenhofer G, Boneberg EM, Legler DF, Plattner H, et al. Preformed reggie/flotillin caps: stable priming platforms for macrodomain assembly in T cells. FASEB J. 2006 Apr;20(6):711-3.

[30] Neumann-Giesen C, Fernow I, Amaddii M, Tikkanen R. Role of EGF-induced tyrosine phosphorylation of reggie-1/flotillin-2 in cell spreading and signaling to the actin cytoskeleton. J Cell Sci. 2007 Feb 1;120(Pt 3):395-406.

[31] Liu J, Deyoung SM, Zhang M, Dold LH, Saltiel AR. The stomatin/ prohibitin/ flotillin/ HflK/C domain of flotillin-1 contains distinct sequences that direct plasma membrane localization and protein interactions in 3T3-L1 adipocytes. J Biol Chem. 2005 Apr 22;280(16):16125-34.

[32] Rajendran L, Beckmann J, Magenau A, Boneberg EM, Gaus K, Viola A, et al. Flotillins are involved in the polarization of primitive and mature hematopoietic cells. PLoS One. 2009;4(12):e8290.

[33] Rajendran L, Masilamani M, Solomon S, Tikkanen R, Stuermer CA, Plattner H, et al. Asymmetric localization of flotillins/reggies in preassembled platforms confers inherent polarity to hematopoietic cells. Proc Natl Acad Sci U S A. 2003 Jul 8;100(14):8241-6.

[34] Frick M, Bright NA, Riento K, Bray A, Merrified C, Nichols BJ. Coassembly of flotillins induces formation of membrane microdomains, membrane curvature, and vesicle budding. Curr Biol. 2007 Jul 3;17(13):1151-6.

[35] Hansen CG, Nichols BJ. Molecular mechanisms of clathrin-independent endocytosis. J Cell Sci. 2009 Jun 1;122(Pt 11):1713-21.

[36] Langhorst MF, Solis GP, Hannbeck S, Plattner H, Stuermer CA. Linking membrane microdomains to the cytoskeleton: regulation of the lateral mobility of reggie-1/flotillin-2 by interaction with actin. FEBS Lett. 2007 Oct 2;581(24):4697-703.

[37] Affentranger S, Martinelli S, Hahn J, Rossy J, Niggli V. Dynamic reorganization of flotillins in chemokine-stimulated human T-lymphocytes. BMC Cell Biol. 2011;12:28.

[38] de Gassart A, Geminard C, Fevrier B, Raposo G, Vidal M. Lipid raft-associated protein sorting in exosomes. Blood. 2003 Dec 15;102(13):4336-44.

[39] Staubach S, Razawi H, Hanisch FG. Proteomics of MUC1-containing lipid rafts from plasma membranes and exosomes of human breast carcinoma cells MCF-7. Proteomics. 2009 May;9(10):2820-35.

[40] Street JM, Barran PE, Mackay CL, Weidt S, Balmforth C, Walsh TS, et al. Identification and proteomic profiling of exosomes in human cerebrospinal fluid. J Transl Med. 2012;10:5.

[41] Strauss K, Goebel C, Runz H, Mobius W, Weiss S, Feussner I, et al. Exosome secretion ameliorates lysosomal storage of cholesterol in Niemann-Pick type C disease. J Biol Chem. 2010 Aug 20;285(34):26279-88.

[42] Dermine JF, Duclos S, Garin J, St-Louis F, Rea S, Parton RG, et al. Flotillin-1-enriched lipid raft domains accumulate on maturing phagosomes. J Biol Chem. 2001 May 25;276(21):18507-12.

[43] Santamaria A, Castellanos E, Gomez V, Benedit P, Renau-Piqueras J, Morote J, et al. PTOV1 enables the nuclear translocation and mitogenic activity of flotillin-1, a major protein of lipid rafts. Mol Cell Biol. 2005 Mar;25(5):1900-11.

[44] Gkantiragas I, Brugger B, Stuven E, Kaloyanova D, Li XY, Lohr K, et al. Sphingomyelin-enriched microdomains at the Golgi complex. Mol Biol Cell. 2001 Jun;12(6):1819-33.

[45] Lang DM, Lommel S, Jung M, Ankerhold R, Petrausch B, Laessing U, et al. Identification of reggie-1 and reggie-2 as plasmamembrane-associated proteins which cocluster with activated GPI-anchored cell adhesion molecules in non-caveolar micropatches in neurons. J Neurobiol. 1998 Dec;37(4):502-23.

[46] Pust S, Dyve AB, Torgersen ML, van Deurs B, Sandvig K. Interplay between toxin transport and flotillin localization. PLoS One. 2010;5(1):e8844.

[47] Kokubo H, Helms JB, Ohno-Iwashita Y, Shimada Y, Horikoshi Y, Yamaguchi H. Ultrastructural localization of flotillin-1 to cholesterol-rich membrane microdomains, rafts, in rat brain tissue. Brain Res. 2003 Mar 7;965(1-2):83-90.

[48] Carcea I, Ma'ayan A, Mesias R, Sepulveda B, Salton SR, Benson DL. Flotillin-mediated endocytic events dictate cell type-specific responses to semaphorin 3A. J Neurosci. 2010 Nov 10;30(45):15317-29.

[49] Riento K, Frick M, Schafer I, Nichols BJ. Endocytosis of flotillin-1 and flotillin-2 is regulated by Fyn kinase. J Cell Sci. 2009 Apr 1;122(Pt 7):912-8.

[50] Resh MD. Fatty acylation of proteins: new insights into membrane targeting of myristoylated and palmitoylated proteins. Biochim Biophys Acta. 1999 Aug 12;1451(1):1-16.

[51] Johnson DR, Bhatnagar RS, Knoll LJ, Gordon JI. Genetic and biochemical studies of protein N-myristoylation. Annu Rev Biochem. 1994;63:869-914.

[52] Dietrich LE, Ungermann C. On the mechanism of protein palmitoylation. EMBO Rep. 2004 Nov;5(11):1053-7.

[53] Li Y, Martin BR, Cravatt BF, Hofmann SL. DHHC5 protein palmitoylates flotillin-2 and is rapidly degraded on induction of neuronal differentiation in cultured cells. J Biol Chem. 2012 Jan 2;287(1):523-30.

[54] Koegl M, Zlatkine P, Ley SC, Courtneidge SA, Magee AI. Palmitoylation of multiple Src-family kinases at a homologous N-terminal motif. Biochem J. 1994 Nov 1;303 (Pt 3):749-53.

[55] Ali MH, Imperiali B. Protein oligomerization: how and why. Bioorg Med Chem. 2005 Sep 1;13(17):5013-20.

[56] Babuke T, Ruonala M, Meister M, Amaddii M, Genzler C, Esposito A, et al. Hetero-oligomerization of reggie-1/flotillin-2 and reggie-2/flotillin-1 is required for their endocytosis. Cell Signal. 2009 Aug;21(8):1287-97.

[57] Langhorst MF, Jaeger FA, Mueller S, Sven Hartmann L, Luxenhofer G, Stuermer CA. Reggies/flotillins regulate cytoskeletal remodeling during neuronal differentiation via CAP/ponsin and Rho GTPases. Eur J Cell Biol. 2008 Dec;87(12):921-31.

[58] Haglund K, Ivankovic-Dikic I, Shimokawa N, Kruh GD, Dikic I. Recruitment of Pyk2 and Cbl to lipid rafts mediates signals important for actin reorganization in growing neurites. J Cell Sci. 2004 May 15;117(Pt 12):2557-68.

[59] Ludwig A, Otto GP, Riento K, Hams E, Fallon PG, Nichols BJ. Flotillin microdomains interact with the cortical cytoskeleton to control uropod formation and neutrophil recruitment. J Cell Biol. 2011 Nov 15;191(4):771-81.

[60] Munderloh C, Solis GP, Bodrikov V, Jaeger FA, Wiechers M, Malaga-Trillo E, et al. Reggies/flotillins regulate retinal axon regeneration in the zebrafish optic nerve and differentiation of hippocampal and N2a neurons. J Neurosci. 2009 May 20;29(20):6607-15.

[61] von Philipsborn AC, Ferrer-Vaquer A, Rivera-Milla E, Stuermer CA, Malaga-Trillo E. Restricted expression of reggie genes and proteins during early zebrafish development. J Comp Neurol. 2005 Feb 14;482(3):257-72.

[62] Stuermer CA. Microdomain-forming proteins and the role of the reggies/flotillins during axon regeneration in zebrafish. Biochim Biophys Acta. 2011 Mar;1812(3):415-22.

[63] Baumann CA, Ribon V, Kanzaki M, Thurmond DC, Mora S, Shigematsu S, et al. CAP defines a second signalling pathway required for insulin-stimulated glucose transport. Nature. 2000 Sep 14;407(6801):202-7.

[64] Katanaev VL, Solis GP, Hausmann G, Buestorf S, Katanayeva N, Schrock Y, et al. Reggie-1/flotillin-2 promotes secretion of the long-range signalling forms of Wingless and Hedgehog in Drosophila. Embo J. 2008 Feb 6;27(3):509-21.

[65] Limpert AS, Karlo JC, Landreth GE. Nerve growth factor stimulates the concentration of TrkA within lipid rafts and extracellular signal-regulated kinase activation through c-Cbl-associated protein. Mol Cell Biol. 2007 Aug;27(16):5686-98.

[66] Lin C, Wu Z, Lin X, Yu C, Shi T, Zeng Y, et al. Knockdown of FLOT1 impairs cell proliferation and tumorigenicity in breast cancer through upregulation of FOXO3a. Clin Cancer Res. 2011 May 15;17(10):3089-99.

[67] Sugawara Y, Nishii H, Takahashi T, Yamauchi J, Mizuno N, Tago K, et al. The lipid raft proteins flotillins/reggies interact with Galphaq and are involved in Gq-mediated p38 mitogen-activated protein kinase activation through tyrosine kinase. Cell Signal. 2007 Jun;19(6):1301-8.

[68] Helms JB, Zurzolo C. Lipids as targeting signals: lipid rafts and intracellular trafficking. Traffic. 2004 Apr;5(4):247-54.

[69] McMahon HT, Boucrot E. Molecular mechanism and physiological functions of clathrin-mediated endocytosis. Nat Rev Mol Cell Biol. 2011 Aug;12(8):517-33.

[70] Staubach S, Hanisch FG. Lipid rafts: signaling and sorting platforms of cells and their roles in cancer. Expert Rev Proteomics. 2011 Apr;8(2):263-77.

[71] Otto GP, Nichols BJ. The roles of flotillin microdomains--endocytosis and beyond. J Cell Sci. 2011 Dec 1;124(Pt 23):3933-40.

[72] Ait-Slimane T, Galmes R, Trugnan G, Maurice M. Basolateral internalization of GPI-anchored proteins occurs via a clathrin-independent flotillin-dependent pathway in polarized hepatic cells. Mol Biol Cell. 2009 Sep;20(17):3792-800.

[73] Blanchet MH, Le Good JA, Mesnard D, Oorschot V, Baflast S, Minchiotti G, et al. Cripto recruits Furin and PACE4 and controls Nodal trafficking during proteolytic maturation. EMBO J. 2008 Oct 8;27(19):2580-91.

[74] Cremona ML, Matthies HJ, Pau K, Bowton E, Speed N, Lute BJ, et al. Flotillin-1 is essential for PKC-triggered endocytosis and membrane microdomain localization of DAT. Nat Neurosci. 2011 Apr;14(4):469-77.

[75] Schneider A, Rajendran L, Honsho M, Gralle M, Donnert G, Wouters F, et al. Flotillin-dependent clustering of the amyloid precursor protein regulates its endocytosis and amyloidogenic processing in neurons. J Neurosci. 2008 Mar 12;28(11):2874-82.

[76] Ge L, Qi W, Wang LJ, Miao HH, Qu YX, Li BL, et al. Flotillins play an essential role in Niemann-Pick C1-like 1-mediated cholesterol uptake. Proc Natl Acad Sci U S A. 2011 Jan 11;108(2):551-6.

[77] Orth JD, Krueger EW, Weller SG, McNiven MA. A novel endocytic mechanism of epidermal growth factor receptor sequestration and internalization. Cancer Res. 2006 Apr 1;66(7):3603-10.

[78] Sigismund S, Argenzio E, Tosoni D, Cavallaro E, Polo S, Di Fiore PP. Clathrin-mediated internalization is essential for sustained EGFR signaling but dispensable for degradation. Dev Cell. 2008 Aug;15(2):209-19.

[79] Solis GP, Schrock Y, Hulsbusch N, Wiechers M, Plattner H, Stuermer CA. Reggies/Flotillins regulate E-cadherin-mediated cell contact formation by affecting EGFR trafficking. Mol Biol Cell. 2012 Mar 21.

[80] Wiley HS, Cunningham DD. The endocytotic rate constant. A cellular parameter for quantitating receptor-mediated endocytosis. J Biol Chem. 1982 Apr 25;257(8):4222-9.

[81] Masui H, Castro L, Mendelsohn J. Consumption of EGF by A431 cells: evidence for receptor recycling. J Cell Biol. 1993 Jan;120(1):85-93.

[82] Schroeder WT, Stewart-Galetka S, Mandavilli S, Parry DA, Goldsmith L, Duvic M. Cloning and characterization of a novel epidermal cell surface antigen (ESA). J Biol Chem. 1994 Aug 5;269(31):19983-91.

[83] Hoehne M, de Couet HG, Stuermer CA, Fischbach KF. Loss- and gain-of-function analysis of the lipid raft proteins Reggie/Flotillin in Drosophila: they are posttranslationally regulated, and misexpression interferes with wing and eye development. Mol Cell Neurosci. 2005 Nov;30(3):326-38.

[84] Yap AS, Brieher WM, Gumbiner BM. Molecular and functional analysis of cadherin-based adherens junctions. Annu Rev Cell Dev Biol. 1997;13:119-46.

[85] Bodrikov V, Solis GP, Stuermer CA. Prion protein promotes growth cone development through reggie/flotillin-dependent N-cadherin trafficking. J Neurosci. 2011 Dec 7;31(49):18013-25.

[86] Pommereit D, Wouters FS. An NGF-induced Exo70-TC10 complex locally antagonises Cdc42-mediated activation of N-WASP to modulate neurite outgrowth. J Cell Sci. 2007 Aug 1;120(Pt 15):2694-705.

[87] Taulet N, Comunale F, Favard C, Charrasse S, Bodin S, Gauthier-Rouviere C. N-cadherin/p120 catenin association at cell-cell contacts occurs in cholesterol-rich membrane domains and is required for RhoA activation and myogenesis. J Biol Chem. 2009 Aug 21;284(34):23137-45.

[88] Chartier NT, Laine MG, Ducarouge B, Oddou C, Bonaz B, Albiges-Rizo C, et al. Enterocytic differentiation is modulated by lipid rafts-dependent assembly of adherens junctions. Exp Cell Res. 2011 Jun 10;317(10):1422-36.

[89] Tomasovic A, Traub S, Tikkanen R. Molecular Networks in FGF Signaling: Flotillin-1 and Cbl-Associated Protein Compete for the Binding to Fibroblast Growth Factor Receptor Substrate 2. PLoS ONE. 2012;7(1): e29739. doi:10.1371.

[90] Kato N, Nakanishi M, Hirashima N. Flotillin-1 regulates IgE receptor-mediated signaling in rat basophilic leukemia (RBL-2H3) cells. J Immunol. 2006 Jul 1;177(1):147-54.

[91] Gotoh N, Manova K, Tanaka S, Murohashi M, Hadari Y, Lee A, et al. The docking protein FRS2alpha is an essential component of multiple fibroblast growth factor responses during early mouse development. Mol Cell Biol. 2005 May;25(10):4105-16.

[92] Lax I, Wong A, Lamothe B, Lee A, Frost A, Hawes J, et al. The docking protein FRS2alpha controls a MAP kinase-mediated negative feedback mechanism for signaling by FGF receptors. Mol Cell. 2002 Oct;10(4):709-19.

[93] Wu Y, Chen Z, Ullrich A. EGFR and FGFR signaling through FRS2 is subject to negative feedback control by ERK1/2. Biol Chem. 2003 Aug;384(8):1215-26.

[94] Hofman EG, Ruonala MO, Bader AN, van den Heuvel D, Voortman J, Roovers RC, et al. EGF induces coalescence of different lipid rafts. J Cell Sci. 2008 Aug 1;121(Pt 15):2519-28.

[95] Ueda Y, Hirai S, Osada S, Suzuki A, Mizuno K, Ohno S. Protein kinase C activates the MEK-ERK pathway in a manner independent of Ras and dependent on Raf. J Biol Chem. 1996 Sep 20;271(38):23512-9.

[96] Zou Y, Komuro I, Yamazaki T, Aikawa R, Kudoh S, Shiojima I, et al. Protein kinase C, but not tyrosine kinases or Ras, plays a critical role in angiotensin II-induced activation of Raf-1 kinase and extracellular signal-regulated protein kinases in cardiac myocytes. J Biol Chem. 1996 Dec 27;271(52):33592-7.

[97] Kouhara H, Hadari YR, Spivak-Kroizman T, Schilling J, Bar-Sagi D, Lax I, et al. A lipid-anchored Grb2-binding protein that links FGF-receptor activation to the Ras/MAPK signaling pathway. Cell. 1997 May 30;89(5):693-702.

[98] Roskoski R, Jr. The ErbB/HER receptor protein-tyrosine kinases and cancer. Biochem Biophys Res Commun. 2004 Jun 18;319(1):1-11.

[99] Ribon V, Herrera R, Kay BK, Saltiel AR. A role for CAP, a novel, multifunctional Src homology 3 domain-containing protein in formation of actin stress fibers and focal adhesions. J Biol Chem. 1998 Feb 13;273(7):4073-80.

[100] Ribon V, Printen JA, Hoffman NG, Kay BK, Saltiel AR. A novel, multifuntional c-Cbl binding protein in insulin receptor signaling in 3T3-L1 adipocytes. Mol Cell Biol. 1998 Feb;18(2):872-9.

[101] Kimura A, Baumann CA, Chiang SH, Saltiel AR. The sorbin homology domain: a motif for the targeting of proteins to lipid rafts. Proc Natl Acad Sci U S A. 2001 Jul 31;98(16):9098-103.

[102] Zhang M, Kimura A, Saltiel AR. Cloning and characterization of Cbl-associated protein splicing isoforms. Mol Med. 2003 Jan-Feb;9(1-2):18-25.

[103] Fecchi K, Volonte D, Hezel MP, Schmeck K, Galbiati F. Spatial and temporal regulation of GLUT4 translocation by flotillin-1 and caveolin-3 in skeletal muscle cells. FASEB J. 2006 Apr;20(6):705-7.

[104] Huang EJ, Reichardt LF. Trk receptors: roles in neuronal signal transduction. Annu Rev Biochem. 2003;72:609-42.

[105] Kaplan DR, Miller FD. Neurotrophin signal transduction in the nervous system. Curr Opin Neurobiol. 2000 Jun;10(3):381-91.

[106] Delahaye L, Rocchi S, Van Obberghen E. Potential involvement of FRS2 in insulin signaling. Endocrinology. 2000 Feb;141(2):621-8.

[107] Meakin SO, MacDonald JI, Gryz EA, Kubu CJ, Verdi JM. The signaling adapter FRS-2 competes with Shc for binding to the nerve growth factor receptor TrkA. A model for discriminating proliferation and differentiation. J Biol Chem. 1999 Apr 2;274(14):9861-70.

[108] Ong SH, Hadari YR, Gotoh N, Guy GR, Schlessinger J, Lax I. Stimulation of phosphatidylinositol 3-kinase by fibroblast growth factor receptors is mediated by coordinated recruitment of multiple docking proteins. Proc Natl Acad Sci U S A. 2001 May 22;98(11):6074-9.

[109] Saslowsky DE, Cho JA, Chinnapen H, Massol RH, Chinnapen DJ, Wagner JS, et al. Intoxication of zebrafish and mammalian cells by cholera toxin depends on the flotillin/reggie proteins but not Derlin-1 or -2. J Clin Invest. 2010 Dec;120(12):4399-409.

[110] Ludwig A, Otto GP, Riento K, Hams E, Fallon PG, Nichols BJ. Flotillin microdomains interact with the cortical cytoskeleton to control uropod formation and neutrophil recruitment. J Cell Biol. 2010 Nov 15;191(4):771-81.

[111] Rossy J, Schlicht D, Engelhardt B, Niggli V. Flotillins interact with PSGL-1 in neutrophils and, upon stimulation, rapidly organize into membrane domains subsequently accumulating in the uropod. PLoS One. 2009;4(4):e5403.

[112] Even-Ram S, Doyle AD, Conti MA, Matsumoto K, Adelstein RS, Yamada KM. Myosin IIA regulates cell motility and actomyosin-microtubule crosstalk. Nat Cell Biol. 2007 Mar;9(3):299-309.

[113] Shih W, Yamada S. Myosin IIA dependent retrograde flow drives 3D cell migration. Biophys J. 2010 Apr 21;98(8):L29-31.

[114] Rajalingam K, Rudel T. Ras-Raf signaling needs prohibitin. Cell Cycle. 2005 Nov;4(11):1503-5.

[115] Rajalingam K, Wunder C, Brinkmann V, Churin Y, Hekman M, Sievers C, et al. Prohibitin is required for Ras-induced Raf-MEK-ERK activation and epithelial cell migration. Nat Cell Biol. 2005 Aug;7(8):837-43.

[116] Ande SR, Gu Y, Nyomba BL, Mishra S. Insulin induced phosphorylation of prohibitin at tyrosine 114 recruits Shp1. Biochim Biophys Acta. 2009 Aug;1793(8):1372-8.

[117] Han EK, McGonigal T, Butler C, Giranda VL, Luo Y. Characterization of Akt overexpression in MiaPaCa-2 cells: prohibitin is an Akt substrate both in vitro and in cells. Anticancer Res. 2008 Mar-Apr;28(2A):957-63.

[118] Chiu CF, Ho MY, Peng JM, Hung SW, Lee WH, Liang CM, et al. Raf activation by Ras and promotion of cellular metastasis require phosphorylation of prohibitin in the raft domain of the plasma membrane. Oncogene. 2012 Mar 12.

[119] Da Cruz S, Parone PA, Gonzalo P, Bienvenut WV, Tondera D, Jourdain A, et al. SLP-2 interacts with prohibitins in the mitochondrial inner membrane and contributes to their stability. Biochim Biophys Acta. 2008 May;1783(5):904-11.

[120] Hazarika P, McCarty MF, Prieto VG, George S, Babu D, Koul D, et al. Up-regulation of Flotillin-2 is associated with melanoma progression and modulates expression of the thrombin receptor protease activated receptor 1. Cancer Res. 2004 Oct 15;64(20):7361-9.

[121] Doherty SD, Prieto VG, George S, Hazarika P, Duvic M. High flotillin-2 expression is associated with lymph node metastasis and Breslow depth in melanoma. Melanoma Res. 2006 Oct;16(5):461-3.

[122] Sasaki Y, Oshima Y, Koyama R, Maruyama R, Akashi H, Mita H, et al. Identification of Flotillin-2, a Major Protein on Lipid Rafts, as a Novel Target of p53 Family Members. Mol Cancer Res. 2008 Feb 22.

[123] Haass C, Lemere CA, Capell A, Citron M, Seubert P, Schenk D, et al. The Swedish mutation causes early-onset Alzheimer's disease by beta-secretase cleavage within the secretory pathway. Nat Med. 1995 Dec;1(12):1291-6.

[124] Kao SC, Krichevsky AM, Kosik KS, Tsai LH. BACE1 suppression by RNA interference in primary cortical neurons. J Biol Chem. 2004 Jan 16;279(3):1942-9.

[125] Small SA, Gandy S. Sorting through the cell biology of Alzheimer's disease: intracellular pathways to pathogenesis. Neuron. 2006 Oct 5;52(1):15-31.

[126] Tikkanen R, Banning A, Meister M. Trafficking and endocytosis of Alzheimer amyloid precursor protein. in: Dowler BC, editor Endocytosis: Structural Components, Functions and Pathways Nova Publishers pp 1-38. 2010.

[127] Riddell DR, Christie G, Hussain I, Dingwall C. Compartmentalization of beta-secretase (Asp2) into low-buoyant density, noncaveolar lipid rafts. Curr Biol. 2001 Aug 21;11(16):1288-93.

[128] Lee SJ, Liyanage U, Bickel PE, Xia W, Lansbury PT, Jr., Kosik KS. A detergent-insoluble membrane compartment contains A beta in vivo. Nat Med. 1998 Jun;4(6):730-4.

[129] Parkin ET, Hussain I, Karran EH, Turner AJ, Hooper NM. Characterization of detergent-insoluble complexes containing the familial Alzheimer's disease-associated presenilins. J Neurochem. 1999 Apr;72(4):1534-43.

[130] Kokubo H, Lemere CA, Yamaguchi H. Localization of flotillins in human brain and their accumulation with the progression of Alzheimer's disease pathology. Neurosci Lett. 2000 Aug 25;290(2):93-6.

[131] Rajendran L, Knobloch M, Geiger KD, Dienel S, Nitsch R, Simons K, et al. Increased Abeta production leads to intracellular accumulation of Abeta in flotillin-1-positive endosomes. Neurodegener Dis. 2007;4(2-3):164-70.

[132] Kokubo H, Saido TC, Iwata N, Helms JB, Shinohara R, Yamaguchi H. Part of membrane-bound Abeta exists in rafts within senile plaques in Tg2576 mouse brain. Neurobiol Aging. 2005 Apr;26(4):409-18.

[133] Girardot N, Allinquant B, Langui D, Laquerriere A, Dubois B, Hauw JJ, et al. Accumulation of flotillin-1 in tangle-bearing neurones of Alzheimer's disease. Neuropathol Appl Neurobiol. 2003 Oct;29(5):451-61.

[134] Chen TY, Liu PH, Ruan CT, Chiu L, Kung FL. The intracellular domain of amyloid precursor protein interacts with flotillin-1, a lipid raft protein. Biochem Biophys Res Commun. 2006 Mar 31;342(1):266-72.

[135] Shin T, Kim H, Jin JK, Moon C, Ahn M, Tanuma N, et al. Expression of caveolin-1, -2, and -3 in the spinal cords of Lewis rats with experimental autoimmune encephalomyelitis. J Neuroimmunol. 2005 Aug;165(1-2):11-20.

[136] Constantinescu CS, Farooqi N, O'Brien K, Gran B. Experimental autoimmune encephalomyelitis (EAE) as a model for multiple sclerosis (MS). Br J Pharmacol. 2011 Oct;164(4):1079-106.

[137] Kim H, Ahn M, Moon C, Matsumoto Y, Sung Koh C, Shin T. Immunohistochemical study of flotillin-1 in the spinal cord of Lewis rats with experimental autoimmune encephalomyelitis. Brain Res. 2006 Oct 9;1114(1):204-11.

[138] Davie CA. A review of Parkinson's disease. Br Med Bull. 2008;86:109-27.

[139] Jacobowitz DM, Kallarakal AT. Flotillin-1 in the substantia nigra of the Parkinson brain and a predominant localization in catecholaminergic nerves in the rat brain. Neurotox Res. 2004;6(4):245-57.

[140] Manes S, del Real G, Martinez AC. Pathogens: raft hijackers. Nat Rev Immunol. 2003 Jul;3(7):557-68.

[141] Vieira FS, Correa G, Einicker-Lamas M, Coutinho-Silva R. Host-cell lipid rafts: a safe door for micro-organisms? Biol Cell. 2010 Jul;102(7):391-407.

[142] Knodler LA, Vallance BA, Hensel M, Jackel D, Finlay BB, Steele-Mortimer O. Salmonella type III effectors PipB and PipB2 are targeted to detergent-resistant microdomains on internal host cell membranes. Mol Microbiol. 2003 Aug;49(3):685-704.

[143] Seveau S, Bierne H, Giroux S, Prevost MC, Cossart P. Role of lipid rafts in E-cadherin-- and HGF-R/Met--mediated entry of Listeria monocytogenes into host cells. J Cell Biol. 2004 Aug 30;166(5):743-53.

[144] Li Q, Zhang Q, Wang C, Li N, Li J. Invasion of enteropathogenic Escherichia coli into host cells through epithelial tight junctions. FEBS J. 2008 Dec;275(23):6022-32.

[145] Chazal N, Gerlier D. Virus entry, assembly, budding, and membrane rafts. Microbiol Mol Biol Rev. 2003 Jun;67(2):226-37, table of contents.

[146] Cicala C, Arthos J, Selig SM, Dennis G, Jr., Hosack DA, Van Ryk D, et al. HIV envelope induces a cascade of cell signals in non-proliferating target cells that favor virus replication. Proc Natl Acad Sci U S A. 2002 Jul 9;99(14):9380-5.

[147] Carter GC, Bernstone L, Sangani D, Bee JW, Harder T, James W. HIV entry in macrophages is dependent on intact lipid rafts. Virology. 2009 Mar 30;386(1):192-202.

[148] Ohmine S, Sakuma R, Sakuma T, Thatava T, Solis GP, Ikeda Y. Cytoplasmic body component TRIM5{alpha} requires lipid-enriched microdomains for efficient HIV-1 restriction. J Biol Chem. 2010 Nov 5;285(45):34508-17.

[149] Nakayama EE, Shioda T. Role of Human TRIM5alpha in Intrinsic Immunity. Front Microbiol. 2012;3:97.

[150] James DJ, Cairns F, Salt IP, Murphy GJ, Dominiczak AF, Connell JM, et al. Skeletal muscle of stroke-prone spontaneously hypertensive rats exhibits reduced insulin-stimulated glucose transport and elevated levels of caveolin and flotillin. Diabetes. 2001 Sep;50(9):2148-56.

[151] Collison M, Glazier AM, Graham D, Morton JJ, Dominiczak MH, Aitman TJ, et al. Cd36 and molecular mechanisms of insulin resistance in the stroke-prone spontaneously hypertensive rat. Diabetes. 2000 Dec;49(12):2222-6.

[152] Ship JA. Diabetes and oral health: an overview. J Am Dent Assoc. 2003 Oct;134 Spec No:4S-10S.

[153] Wang D, Yuan Z, Inoue N, Cho G, Shono M, Ishikawa Y. Abnormal subcellular localization of AQP5 and downregulated AQP5 protein in parotid glands of streptozotocin-induced diabetic rats. Biochim Biophys Acta. 2011 May;1810(5):543-54.

Chapter 4

Modulation of HER2 Tyrosine/Threonine Phosphorylation and Cell Signalling

Yamuna D. Gandaharen, Rachel S. Welt, David Kostyal and Sydney Welt

Additional information is available at the end of the chapter

1. Introduction

Cell-surface transmembrane tyrosine kinase signalling receptors EGFR, HER3 and HER4 are activated by specific ligand binding, reference in [1] which, induces signalling through intra-cellular domain kinase-dependant phosphorylation sites, see [2]. The intensity and specificity of the transmitted signal are modulated by homo and hetero-dimerization, allowing other family members to be recruited and thereby amplifying the signal [3]. HER2, which does not have a defined activating ligand, plays a key role in its capacity to amplify the signal [4] by being the preferred partner for hetero-dimerization. Both HER2 hetero and homo dimers maintain their intracellular kinase and phosphorylation sites in an activated state [3,5,6,]. HER2 hetero and homodimers both maintain their intracellular kinase and phosphorylation sites in an active state [5,6] Dimerization with HER2 increases receptor-ligand affinity and receptor-ligand stability, further enhancing activation [4,7]. In vivo enzymatic processing releases the HER2 extra-cellular domain (95 kD), including the dimerization site [8] from the membrane, but the role this plays in signal modulation remains undefined [9]. HER2-dependant activation signals promote a highly phosphorylated tyrosine state in the intra-cellular domain, which is then recognized by a family of specific cytoplasmic signal transducers [10]. These signal transducers regulate cell proliferation, apoptosis and cell characteristics associated with the transformed state [1,2 5,6]. Mitogen-activated protein kinases (MAPKs), and phosphatidylinositol 3-OH kinase (PI3K- AKT) pathways [11,12] have been particularly well studied as downstream signal pathways of HER2 leading to the concept of an oncogenic unit [13]. Accumulating experimental and clinical data link anti-HER2 treatment to inhibition of these downstream pathways suggesting that they are critical mediators of the activated HER2 state and cell survival [14]. However, whether the trastuzumab mediated inhibition of signalling is due to reduced HER2 phosphorylation/kinase activity, inhibition of HER2 dimerization, decreased HER2 levels, altered metabolic processing of HER2, binding dependant allosteric effects on HER2 or HER2-HER1/3/4 or a combination of these mechanisms is unknown.

In most cases, amplification of HER2 in breast cancer is associated with an amplicon comprising genes mapped to the 17q12-q21 region [15-18] that includes several genes involved with cell proliferation [16], transformation [17], adhesion [16] and chemotherapy sensitivity [18]. Clinically, the amplicon gene copy number is a parameter closely linked to trastuzumab and lapatinib efficacy [19,20].

Experimental evidence now suggests that nuclear localization of growth factors and receptors is not uncommon and also occurs with members of the EGFR family [21]. More recently, specific products derived from the HER2 receptor have been shown to have independent gene regulatory effects by virtue of a nuclear localization motif encoded in the HER2 protein [21,22] and by alternate splicing of the HER2 mRNA [21]. For example, the nuclear localization sequence preferentially targets a HER2 derived peptide to bind to sequences, one of which is in the COX-2 promoter region, leading to up-regulation of COX-2 expression [22]. Whether there are other HER2 derived peptides and the role they may play in the signalling cascade and whether any of the current anti-HER2 treatments modulate these factors remains is an area of current interest.

The original anti-HER2 mouse monoclonal antibody, 4D5, was shown to inhibit HER2 tyrosine phosphorylation [23] and breast cancer cell proliferation [23]. This antibody was humanized through genetic engineering to become trastuzumab [24,25]. A second antibody now in development (pertuzumab) binds to an extra-cellular HER2 epitope in close proximity to the dimerization site and has been shown to block dimerization [26-, 28]. A small molecule HER2/1 tyrosine kinase inhibitor, lapatinib has shown single agent activity in the clinic [20,29,30], providing further evidence that the disruption of the EGFR1/HER2 activation pathway, specifically by blocking phosphorylation [29-32], leads to cell death.

The focus of this investigation was to examine the effects of trastuzumab and lapatinib on breast cancer cell lines containing the HER2 amplicon under conditions of steady state anti-HER2 growth inhibitory effects. The quantifiable end point was the inhibition of proliferation as determined by MTT and Tritiated thymidine incorporation. These assays were used as evidence that the trastuzumab or lapatinib treatments had interrupted signalling pathways. Using immune detection of three distinct HER2 protein sites, four distinct HER2 tyrosine phosphorylation (P-Tyr) sites, a threonine phosphorylation (P-Thr) site and actin, changes in these parameters were monitored during anti-HER2 treatment. The demonstration of the restricted tyrosine kinase inhibitory activity of lapatinib further validated the specificity of the anti-HER2 phospho-tyrosine/threonine (P-Tyr/ P-Thr) antibodies selected for this investigation.

2. Materials and methods

Breast cancer cell lines were obtained from the American Type Culture Collection (ATCC, Manassas, VA) specifically for these experiments and maintained in RPMI 1640 with 50 units penicillin/streptomycin, 2 mM glutamine, 7.5 % fetal bovine serum, and incubated at 37° C in 5% CO_2. The ATCC authenticates and tests cell lines provide for research as per their protocols.

Four drug treatments and a control, done in quadruplicate, were analyzed using Western blot techniques. Cell concentrations were adjusted to 375,000 cells/ml in complete media and 1 ml was added to each well of a 12 well tissue culture plate. Cells were allowed to adhere to the plates for 24 hrs. Drugs or control solutions were added at a final concentration of 100 microgram/ml trastuzumab, (Herceptin) or control rituximab (Rituxin), (Genentech, San Francisco, Ca). Preliminary dose finding experiments were carried out at 2-250 micrograms/ml. 0.1% DMSO and 0.1%DMSO + 10 ηM Lapatinib (Tykerb) (GlaxoSmithKline Research Triangle Park, NC) was also tested following this same procedure .The cells were incubated at 37º C in 5% CO_2 for 8 or 48 hrs. The cells were lysed with either 400 µl of SDS loading buffer (58 mM Tris 6.8, 1.6 % SDS, 6% glycerol, 0.83% BME, 0.002% bromphenol blue) or NP-40 (0.5% NP-40, 50 mM Tris 7.4, 120 mM NaCl, 2.5 mM EGTA). Additional cell growth inhibitory studies were carried out for incubations up to 9 days in 6-well plates, and cells were tested serially every other day by pulse Tritiated thymidine incorporation and MTT assays

For Western blotting, samples were boiled in electrophoresis sample buffer containing 0.0625M Tris-HCl (pH6.8), 10% glycerol, 2% SDS and 5% BME, for 10 minutes then separated on 12% SDS-PAGE gels (mini-protean, Bio-Rad, Richmond, CA) at 100 V until the dye front reached the bottom then transferred overnight (47 mA) to nitrocellulose membranes (Bio-Rad). Membranes were blocked with a 5% w/v solution of non-fat dry milk in TPBS for 2 hours at room temperature (RT). The blots were then probed with primary antibodies according to the respective titers provided by their manufacturers and the membranes were rocked for 2 hours at RT. The blots were washed 1X with TPBS and a 1:10,000 dilution of secondary antibody (either goat anti-rabbit alkaline phosphatase or goat anti mouse alkaline phosphatase (Sigma, ST Louis, MO) was added, and the blots rocked at RT for 1 hour. The blots were washed 3X with TPBS and the membrane developed in the alkaline phosphate substrate NBT/BCIP (Promega Corp., Madison, WI).

Rabbit anti-P-Tyr 877, P-Tyr 1112, P-Tyr 1221/1222, P-Tyr 1248 P-Tyr 877, P-Tyr 1221/1222, and anti-amino acid peptide containing residue 1222, were purchased from Cell Signalling Technology, Inc,(Danvers, MA). Anti-P-Tyr 4G10, was purchased from Millipore, (Billerica, MA). Anti-extracellular HER2, Neu (9G6):sc-08, anti-HER2 P-Thr 686, p-Neu (7F8):sc81508, anti-HER2 P-Tyr 1112, p-Neu (19G5):sc-81528, anti-extracellular HER2, Neu(ER23):sc-74241, were purchased from Santa Cruz Biotechnology (Santa Cruz, CA). Anti-HER2 (AA 7570-987) was purchased from ProMab Biotechnologies, Inc. (Albany, CA). All antibodies were used per the manufacturer's recommendations. Membrane fractions were isolated using the Perfect-Focus Membrane Protein kit, (G Biosciences, Maryland Heights, MO), according to the manufacturer's protocol.

Human serum samples from patients and normal volunteers were collected under an Institutional Review Board approved research protocol and stored in our serum bank of de-identified serum specimens, stored at -78 degrees Celsius.

3. Results

3.1. Cell growth-inhibitory and viability assays

To define the experimental time points where anti-HER2 therapy modulation of downstream signalling could be demonstrated, breast cancer cell lines were tested for trastuzumab/lapatinib growth inhibitory effects by Tritiated thymidine incorporation and MTT assays. Three HER2 amplicon-containing cell lines were examined, BT-474, SK-BR-3, MDA-MB-453, and amplicon (-) MCF7 served as a control [29,31]. Cells were treated in log phase growth under conditions where factors such as confluence inhibition were minimized. Antibody controls included isotype matched human/mouse IgG1 chimeric (rituximab) and fully humanized IgG1 antibody (bevacizumab). In addition, IgG1 cetuximab, which blocks EGFR1 signalling, was used as an additional control; as in a separate set of experiments, we confirmed that bit did not induce a proliferation or inhibitory state in these cell lines, in agreement with reports by others, nor did we find any effect on HER2 tyrosine phosphorylation [29]. Trastuzumab was tested at a continuous cell exposure dose of 100 micrograms/ml with select experiments carried out at 2- 250 micrograms/ml. Cell cultures were assayed at various time points up to 9 days of treatment. Results of cumulative and pulsed-chased Tritiated thymidine incorporation and MTT assays were in agreement, demonstrating a reproducible inhibitory effect of trastuzumab on cell growth and division. This is best demonstrated by the cell line BT-474. Cell lines SK-Br-3 and MDA-MB-453 exhibited lesser effects and no effect was seen on HER2 amplicon negative control MCF7 cells. These results are consistent with previous reports [29,31]. The Tritiated thymidine incorporation experiments suggested that a trastuzumab-mediated continuous cell-growth inhibitory state is established throughout the incubation period by demonstrating reduced incorporation per cell at each time point tested in the trastuzumab treated cultures (Table 1). Significant cell death was not observed with trastuzumab up to the 9-day time point. Of note was that the maximum cell-growth inhibition by both MTT and Tritiated thymidine incorporation assays was observed at 5 micrograms/ml of trastuzumab, thus the assays described below were carried out at 20-50 times the active biological dose. The 100 micrograms/ml dose was selected as it most closely matches doses achieved in patients.

Effective doses for lapatinib in the cell culture assays ranged from 1 nanogram/ml to 1 microgram/ml. In contrast to results with trastuzumab, lapatinib induced significant cell death by day five (> 90%) under conditions of continued cell exposure at a constant drug dose of ≥10 nanograms. Lapatinib cell cultures beyond five days had no measurable MTT signal, lack of Tritiated thymidine incorporation (< 1% initial values) and few viable adherent cells by microscopic examination. Experiments with lapatinib included a dimethyl sulfoxide (DMSO) control, as it was the solvent used for this lipophilic drug. Control DMSO had a small effect on growth kinetics but no cell death was observed. In order to examine the comparative biologic/phosphorylation inhibition effects of these two agents, time points of 8 hours and 48 hours were selected for further analysis. The critical feature of these two time points is that cell viability appeared high in both trastuzumab and lapatinib cultures as demonstrated by the MTT assay, Tritiated thymidine incorporation, protein levels of HER2 and actin in cell cultures as assessed by Western blots, and the continued adherence of cells up to 48 hours as determined by microscopic examination. These experiments would therefore allow for the examination of molecular events prior to cell death, distinguishing between events that may be related to induction of cell death versus changes resulting from cell death.

TRASTUZUMAB	DAYS	1	3	5	7	9
% ^{3}H-thymidine incorporation	Cell lines					
	BT474	76.3	81.5	76.7	68.5	66.2
	SK-BR-3	84.4	80.2	80.4	77.6	65.1
	MDA-MB-543	81.4	84.3	78.3	73.9	67.7
	MCF-7	98.7	97.9	101.6	99.0	100.7
% MTT						
	BT474	99.3	95.4	87.6	62.1	45.4
	SK-BR-3	97.7	93.6	75.5	70.1	65.3
	MDA-MB-543	99.7	97.3	90.3	82.1	71.2
	MCF7	97.7	99.8	102.2	100.6	101.0
LAPATINIB	DAYS	1	3	5	7	9
% ^{3}H-thymidine incorporation	Cell lines					
	BT474	80.3	77.2	6.6	1.1	----
	SK-BR-3	84.7	82.1	8.8	----	----
	MDA-MB-543	88.7	82.0	8.4	----	----
	MCF-7	102.0	101.2	96.7	98.5	97.9
% MTT						
	BT474	97.6	88.4	4.7	----	----
	SK-BR-3	99.9	96.8	14.3	----	----
	MDA-MB-543	100.5	97.9	13.6	----	----
	MCF-7	104.1	97.9	97.6	101.7	100.6

Values are expressed as %-treated cells/controls for each time point. Data shown represents experiments carried out with trastuzumab 100 micrograms/ml and lapatinib 10 nanograms/ml While the MTT assay represents relative total viable cell counts of treated cells compared to controls, the ^{3}H-thymidine counts/minute (cpm) were corrected for number of viable cells using the MTT assay data (cpm treated/cpm control/MTT treated/MTT control). The paired data sets were evaluated by Student's t-test for significance and those values achieving statistical significance are **bolded**. ---- indicates no measurable signal.

Table 1. Relative growth inhibitory values

3.2. Western blot experiments

3.2.1. HER2 protein levels

Protein level changes for HER2 and actin, associated with trastuzumab or lapatinib treatment at two selected treatment time points (8 and 48 h) were analyzed by Western blot analysis. Three specific polyclonal antibody sera detecting HER2 were tested, one recognizing epitopes in the ectodomain (amino acid 182-373), and two detecting epitopes in the cytoplasmic domains (amino acids 750-987 and a determinant in proximity to the Tyr 1222 amino acid). These experiments were carried out at least four times each and representative panels of a single experiment are shown in Figure 1. The three controls shown here were cells grown in media alone, cells grown in the presence of rituximab, and cells grown in DMSO. All controls gave very similar bands for the three anti-HER2 antibody sera. Small changes are inHER2 levels are first detected at 48 hours in cell line SK-Br-3 treated with lapatinib, while actin levels remain unchanged at this time point. This finding is consistent in each of the four separate experiments and confirms the increase in cell surface HER2 levels associated with lapatinib treatment as previously reported for cell line SK-Br-3 [31]. This time point is just prior to the onset of events associated with apoptosis and may represent a consequence of cellular pathway disruption, as increased HER2 mRNA levels were not observed. In separate experiments, cells grown in the presence of cetuximab or bevacizumab (not shown) resulted in western blots identical to controls indicating a lack of modulation of HER2 protein expression by anti-EGFR or VEGF therapy in these cell lines. These results are consistent with our previously reported analysis of EGFR, HER2, HER3 and HER4 mRNA levels by gene expression mRNA array analysis of mRNA extracted from these breast cancer cell lines treated with trastuzumab compared to these controls [30]. These parallel experiments used extracts from these breast cancer cell lines treated with trastuzumab or cetuximab, or control antibodies, under these same defined conditions and time points.

No difference in actin levels was noted in each set of these experiments. As the cell count number in the first 48 hours showed <10% difference, the amount of cell extract applied to the gel while corrected for cell number at each time point had very little variance. Thus, these results are consistent with those of the Tritiated thymidine and MTT assays indicating the minor growth inhibitory effects of these agents up to 48 hours. These finding also support the contention that biologic events occurring up to 48 hours of treatment precede cell death events and are possible activators of critical pathways leading to cell death.

3.2.2. Tyrosine and threonine phosphorylation levels in the control cultures:

Four site specific anti-P-Tyr antibodies (Tyr 877, 1112, 1222 and 1248) and one anti-P-Thr (Thr 686) were tested on cell extracts collected at specific timed intervals (Figure 1). The controls showed consistent pattern of staining at the two time points examined for each cell line, but different patterns of relative intensity for each tyrosine or threonine were observed for each cell line. These results suggest that each cell line has a distinct net P-Tyr signalling oncogenic unit, rather than a HER2 specific net down-stream signalling motif.

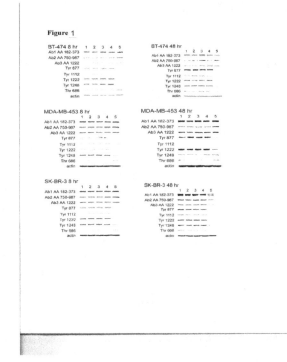

Figure 1. Shows Western blots bands of HER2, phosphorylated tyrosines/threonines, and actin from representative experiments of three cell lines growing in the presence of trastuzumab or lapatinib, or their controls, tested at 8 hours and 48 hours after drug administration. Primary antibodies were used at 1:5,000 for anti-P-Tyr/Thr and anti-HER2 antibodies and 1:10,000 for anti-actin antibodies and secondary antibodies. Labelled lanes show trastuzumab-treated cells in *lane* 2 with its controls in *lane* 1 (media alone) and 3 (control antibody). *Lane* 5 shows lapatinib-treated cells and the control grown in lapatinib solvent, (DMSO) in *lane* 4. Blots show the p185 band with secondary reagent as shown, and actin is presented as proof of viability and equivalent cell number applied to each *lane*. These gels demonstrate the contrast in phosphorylated tyrosines detected comparing lapatinib to trastuzumab and confirm the TKI activity of lapatinib. At 8 hours, lapatinib almost completely blocked tyrosine phosphorylation with virtually no discernible difference in the in the HER2 protein levels. The 48 hour time point was selected as it represented the last time point tested prior to lapatinib-induced cell death, which was almost complete by day 6.

Compared to the media controls, the DMSO controls did not differ in phosphorylation levels of these specific P-Tyr or P-Thr. While the intensity of the bands were similar for each cell line at each of the two time points, exceptions were noted for Tyr 877 and cell lines BT-474 and MDA MD 453, where phosphorylation is clearly increased from the 8 hour to the 48 hour time point. Similarly, P-Tyr 1222 also demonstrated increased intensity from 8 to 48 hours in cell line MDA-MB-453. Whether these findings suggest that the cells are not completely recruited into log phase growth at the 8-hour time point or cell-cell contact plays a role is unclear. The actin and HER2 bands strongly suggest equivalence of cell number applied to each lane at each of these time points.

3.2.3. Tyrosine and threonine phosphorylation levels in trastuzumab and lapatinib treated cells

No discernable change in phosphorylation of the P-Tyr or P-Thr was detected for trastuzumab-treated cells despite ongoing inhibition of cell growth at these time points. The unexpected lack of trastuzumab inhibition of tyrosine phosphorylation prompted experiments that replaced fetal calf serum (FCS) with human serum to determine if a component of FCS was responsible for inhibiting the full trastuzumab activity. Cells grown in media with 10% human serum from breast cancer patients or normal females, pre or postmenopausal, or male serum did not affect these findings (data not shown). Doses of trastuzumab up to 250 micrograms/ml did also did not change the patterns of phosphorylation. In contrast to the results demonstrating constant HER2 protein steady state levels, P-Tyr levels in lapatinib treated cells are markedly reduced (Tyr 1248) or absent (Tyr 687, 1116 and 1223), consistent with its role as a HER2 tyrosine kinase inhibitor (Figure 1). In addition, the P-Thr level is not affected by lapatinib exposure, demonstrating the tyrosine backbone specificity of lapatinib. As the lapatinib Western blots demonstrate, HER2 and actin protein levels are maintained through 48 hours, these results strongly suggests that changes in detected phosphorylation levels are due to kinase inhibition rather than the substantial in HER2 protein or cell death. We suggest that these early findings are related to the lapatinib mechanism of cell death that is apparent 2-3 days later.

3.3. MRNA expression patterns in trastuzumab treated breast cancer cells

3.3.1. Detection of a small molecular weight HER2-like product

Cell fractionation experiments were carried out to determine whether HER2 derived products with nuclear localization motifs could be identified by Western Blot analysis and to determine if trastuzumab or lapatinib altered their molecular processing mechanism. Differential generation of enzymatically derived nuclear localizing products could be responsible for trastuzumab mediated trans-membrane growth inhibitory effects. The results indicate that these techniques were not sensitive enough to detect, or were not of the correct immune-specificity to detect HER2 derived product in concentrates of nuclear or cytoplasmic cell fractions (data not shown).

However, using the anti-HER2 external domain antibody (AA 182-373), we detected a small molecular weight (20 kD) external domain-like HER2-derived product associated with the cell (Figure 2). This peptide segment is too far from the hydrophobic transmembrane segment to contain a transmembrane anchoring component to explain its association with the cell pellet, as it is too small (approximately 180 amino-acids in length) to extend from position 373 to the transmembrane domain. This peptide was not detected in supernatant or concentrates of supernatants. Cells exposed to acid conditions (pH 4.0, 0.5 M acetate, 0.15 N NaCl) to elute this 20 kD segment from the cell surface demonstrated that the 20 kD peptide could not be eluted under these conditions and was still found associated and concentrated in the cell pellet fraction (figure 2). Because concentrates of cytoplasmic fractions and nuclear fractions did not demonstrate the presence of this 20 kD peptide segment, a Triton X-114 fraction of the cell membrane was assayed and found to have high levels of this product (Figure 2). These results suggest that this product is derived from the extra-cellular domain of HER2 by enzymatic processing and binds avidly to a cell membrane-anchored determinant or is internalized into a sub-membrane compartment, co-purifying with the membrane fraction, or that it is a membrane protein cross reactive with antibody to extracellular domain (AA 183-373). The significance of this finding is that this peptide appears to be inhibited by lapatinib treatment while the intact HER2 levels remain constant and expression of this protein is unaffected by trastuzumab treatments (Figure 2). Thus, this small molecular weight product may be related to the unphosphorylated state of HER2.

4. Discussion

HER2 transmembrane activation signals promote a high level of selected intracellular domain P-Tyr sites, which are then recognized by specific cytoplasmic signal transducer molecules [10]. Thus, HER2 acts as an oncogenic unit transmitting cell survival, growth, and cell proliferation signals to critical intracellular pathways such as Mitogen-activated protein kinases (MAPKs), AKT and phoshatidylinositol 3-OH kinase (PI3K) pathways[11,12,29]. Here, we show that the relative level of specific P-Tyr is different for each HER2 amplicon-positive cell line examined, and thus, each oncogenic unit may be distinct in each tumour [13]r. Linkage of anti-HER2 treatment to the inhibition of these downstream pathways is well established, suggesting that these specific P-Tyr are critical mediators of the downstream signalling effects of the activated HER2 state and cell survival [7,10-12,29,30]. The molecular mechanism by which trastuzumab, a humanized IgG1, modulates the intracellular tyrosine kinase site from its extra-cellular binding site has remained unanswered. For example, cetuximab binds to, or near, the ligand binding site of EGFR thus acting as an antagonist [1,2,30-32]. Defining the molecular mechanism by which trastuzumab mediates its anti-HER2 effects could reveal a general non-ligand site-specific strategy for development of receptor inhibitory monoclonal antibodies in other receptor kinase systems.

The primary focus of this investigation was to determine the key molecular mechanism(s) governing trastuzumab's transmembrane modulation of downstream effects. Here, we show that trastuzumab binding does not appear to alter the net steady state levels of p185 HER2 protein as detected by all three of the anti-HER2 antibodies tested. Previous gene mRNA expression array findings demonstrate that no significant change in mRNA levels for HER2 or the other members of the ERBb family were induced by trastuzumab binding [33]. Taken together, these results strongly argue against an antibody-dependent alteration in natural turnover rate or internalization kinetics of HER2 [34]. We conclude that in the absence of demonstrable modulation of HER2 membrane mass on the cell surface, HER2 protein synthesis, and thus, alteration of kinetics of internalization or HER2 metabolism, net membrane HER2 levels could not be implicated as a mechanism to down-regulate HER2 signalling. Despite these observations for HER2, our gene mRNA expression array data did suggest numerous intracellular pathway modulations mediated by trastuzumab binding to the target cells examined here [33]. These findings confirm transmembrane signal transduction modulation directed and specific to trastuzumab binding [33-35]. For example, we found that, the ubiquitin pathway is up-regulated upon trastuzumab binding to HER2 [35], as has been reported by others through ligase activity at tyrosine 1112 [35,36].

Trastuzumab has not been definitively shown to retain the tyrosine kinase inhibitory activity of the parent 4D5 mouse monoclonal antibody [23]. Here we show that the level of HER2-P-Tyr/Thr corrected to cell number, actin levels, and the steady-state levels of intact extractable HER2, is not altered by exposure to trastuzumab. The finding that both P-Tyr levels and HER2 levels in each of these assays reveal no change further supports the conclusions derived from the gene mRNA expression array data [33]. Thus, examination of HER2 steady-state protein levels and key phosphorylation sites [10] did not reveal the molecular perturbation responsible for well-documented trastuzumab inhibition of cell proliferation and downstream pathways.

In contrast to trastuzumab, lapatinib treatment almost completely abrogates the HER2-P-Tyr as a signalling element while leaving HER2 protein levels unchanged. Thus, lapatinib acts through the blockage of tyrosine phosphorylation rather than the modulation of internalization or metabolism of the HER2 protein. We further show a sequential relationship between lapatinib-induced blockage of HER2-P-Tyr (8 and 48 hours) and induction of cell death at day 5-6 of drug exposure. At this late time point, MTT, Tritiated thymidine, actin levels and microscopy all support a lapatinib induction of cell lyses.

The inhibition of cell growth was used to define the conditions of ongoing drug-induced biological effects. The cell growth experiments described here are in agreement with similar reports by others [24,31]. Trastuzumab induced slower growth but did not suppress net increases in cell numbers over time. This *in vitro* effect of trastuzumab parallels the clinical observation that a prolonged stable disease is a common outcome of treatment [37].

Treatment of amplicon-positive breast cancer cells with cetuximab confirms reports by others that anti-EGFR therapy does not inhibit proliferation, induce cell death, nor inhibit HER2 phosphorylation in these cell lines [31]. Furthermore, we did not find that combinations of cetuximab and trastuzumab augment inhibition of cell growth or promote apoptotic events [31]. Cell growth inhibition and phosphorylation experiments with cetuximab have not been in agreement with results of similar investigations using small molecules with EGFR tyrosine kinase inhibitory activity (TKI) [38,39]. However, drugs such as gefitinib, which are believed to be relatively specific for EGFR, are also found to block HER2 phosphorylation at higher doses [38], complicating the interpretation of its mechanism of action [38]. Whether there is an independent anti-EGFR activity cross-inhibiting HER2 phosphorylation (cross-talk) or whether this observation is a consequence of the direct anti-HER2-phosphorylation activity of gefitinib is unclear. Results of these *in vitro* experiments may mislead investigators into attributing findings to cross-talk between receptor pathways rather than cross-reactivity of the reagent tested. We may over estimate biologic activity when reagents are used at too high of a concentration or too long of an exposure. Results of these experiments may mislead investigators into attributing findings to cross-talk between receptor pathways rather than the cross reactivity of the reagent tested. Thus, we conclude that results of experiments testing cetuximab's biologic effects more closely represent, and are more specific to, the pure anti-EGFR blockade. We clearly show here, in agreement with previous reports by others that pure anti-EGFR1 inhibition of signalling does not result in cell death or modulation of HER2 phosphorylation levels [29].

These observations suggest that the anti-EGFR1 tyrosine kinase effects of lapatinib are not contributing to its anti-HER2 effects. Thus, the results of the panel of anti-HER2-P-Tyr-antibodies demonstrating almost complete abrogation of HER2 phosphorylation by lapatinib and its associated delayed cell cytotoxicity (day 5-6) establishes a specific and unique link between HER2 tyrosine phosphorylation and cell survival. Importantly, these results also suggest that additional signalling through HER3 and HER4 may not be critical for apoptosis [40], except as it may ultimately be mediated through HER2 tyrosine kinase inhibition (reduction in hetero-dimerization) [40]. Thus, blocking dimerization by itself, which is unlikely to affect steric change in the intracellular domain, will not induce a modulation of P-Tyr levels. Taken together, these data suggest that testing the effect of a specific and robust HER2 tyrosine kinase inhibitory reagent is warranted.

In the clinic, where the anti-EGFR toxicity of lapatinib is dose limiting (diarrhea and skin rash) and prohibits dose escalation, the testing of a small molecule TKI manifesting pure HER2 kinase inhibition would be of great interest. Recently a pan ErbB TKI was tested and shown to be dose limited by the EGFR-mediated bowel toxicity, further supporting the conclusions presented here [41]. A pure anti-HER2 agent would be even more compelling if one could demonstrate upon dose escalation increased tumor cytotoxicity and a lack of other intervening non-HER2 related dose-limiting toxicities. Close examination of many of the TKIs demonstrate widespread promiscuity with many unexpected cross-reactivities [42]. In general, clinical success of a widely reactive kinase inhibitor is not informative as an identifier of the key pathway(s) responsible for mediating anti-tumor effects and often leads to "assumed" mechanisms of activity.

Using the four anti-P-Tyr sera, we could not find any diminution in the phosphorylation levels as assayed by Western blots, in cells treated up to nine days in the presence of trastuzumab. In addition, the anti-phospho-threonine reagent was used as a negative control, and was found not to be blocked by lapatinib demonstrating the specificity of the drug and the kinase activity. Thus, we propose that the kinase inhibition that was originally described by the mouse antibody 4D5 [23] has been lost in the humanization process [25]. The abundance of well-documented evidence that inhibition of specific downstream pathways is a consequence of trastuzumab-cell binding remains unexplained.

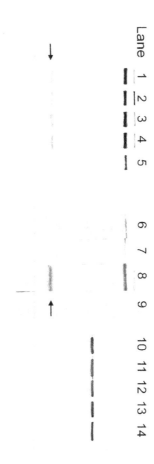

Figure 2. Demonstrates a series of experiments evaluating a 20 kD protein detected by anti-HER2 extracellular domain (AA 182-373) anti-sera at 48 hours. *Lanes 1-5* demonstrate the full Western blot gel of BT-474 (48 hours, from Figure 1) showing two bands: the p185 HER2 protein and the 20 kD protein (arrow on left) *lanes 1-4*. Note that this protein in not detected in *lane 5* which represents lapatinib-treated cells, while the p185 HER2 is mildly reduced in *lane 5*. *Lanes 1- 14* show the corresponding actin control confirming that equal numbers of cell were applied to *lanes 1-5*. Exposing BT-475to cells to an acid wash of pH 4.0 does not elute the protein as *lane 7* is the acid wash and the cell pellet that remains is shown in *lane 6*, has both p185 and the 20 kD protein. Using the Triton X Perfect Focus Membrane Protein kit methodology shows a clearer picture of localization of the 20 kD band with membrane HER2 protein and the 20 kD band (arrow) both in *lane 8* (membrane protein concentrate) and corresponding cell pellet, run in *lane 9*, is negative.

One hypothesis to be considered is that the trastuzumab effect is a weaker version of what is seen with lapatinib and while we cannot demonstrate a small loss of P-Tyr activity in culture, more sensitive techniques might be informative. Alternatively, trastuzumab inhibition may be allosteric or conformational in nature and at the level of the cytoplasmic signal transducers [10] ability to detect the phosphate moieties in the context of an altered peptide conformation [43]. One may further speculate that due to the unique binding epitope of trastuzumab, which is in extra-cellular proximal domain, trans-membrane tertiary structural changes may be induced. Cross-linking HER2 (dimers) at the extracellular proximal domain may disrupt the P-Tyr access to cytoplasmic signal transducers by physical constraints. This hypothesis would suggest that other membrane receptor systems could also be modulated in a similar fashion.

Our results demonstrate the presence of a small 20 kD protein concentrated in the extracted membrane fraction which is immunologically similar to an extra-cellular segment of HER2 protein. This may provide further evidence of specific enzymatic digestion of HER2 into novel biologic agents [44,45]. The low molecular weight of this identified fragment and its association with the membrane fraction raises the possibility that it may function as a ligand for another, as yet unidentified, receptor. Thus, we hypothesize that HER2 may not be a classical receptor as has been suspected by the lack of a ligand binding moiety but rather a pre-ligand entity which requires regulated and specific digestion to be activated. The differential expression of this moiety with lapatinib in contrast to trastuzumab treatment, demonstrates the complexity of this system and the difficulty in attributing a single molecular mechanism in transmembrane signal transduction.

These observations are not informative regarding the precise cytotoxic mechanism of trastuzumab activity in the clinic. Models based on preclinical animal studies favor immune mechanisms as key in mediating tumor shrinkage [46]. Studies of patients treated in the neo-adjuvant setting also support these findings [43]. However the clinical effects of trastuzumab could be divided into two distinct effects, a growth inhibitory effect as seen in vitro and which manifests clinically as long term stable disease and a cytotoxic effect as one might expect from immune mediated cell lysis which results in rapid tumor shrinkage in the clinic, especially as seen in neo-adjuvant studies [47].

In addition to these hypotheses, there is evidence that HER2 induced downstream effects actually amplify pro-inflammatory factors such as COX-2 [22] creating a micro-environment for promoting immune cell mediated tumor cell killing. These data are in agreement with modulation of a prostaglandin pathway identified by gene mRNA expression array experiments [33] in trastuzumab treated cells. Thus downstream events induced by HER2 or regulated by trastuzumab binding, may promote the immune mediated killing of tumor cells by trastuzumab. These findings provide a strong rational for combining other anti-breast cancer cell antibodies with trastuzumab therapy. Furthermore the additional anti-tumor efficacy observed with combinations of trastuzumab and chemotherapy could also be attributable to the known effects of certain chemotherapeutic agents, which render tumor cells more sensitive to immune attack [48]. If this is the case, research should be focused on mechanisms by which chemotherapy may damage tumor cells sufficiently to allow more efficient killing by immune mechanisms rather than focus on apoptotic pathways that enhance chemotherapy cytotoxic effects.

Author details

Yamuna D. Gandaharen and Sydney Welt
Welt Bio-Molecular Pharmaceutical, LLC. Molecular Research, Armonk, NY, U.S.A.

Rachel S. Welt
Department of Biological Sciences, Fordham University, NYC, NY, U.S.A.

David Kostyal
ARD Laboratories, Biologic Research, Akron, OH, U.S.A.

Acknowledgement

Funding and support-PA State Department of Health Project Cure Award SAP # 4100031280 and the Cantwell Foundation Fund of Coudersport Pennsylvania 16901.

Disclosure statement

The authors declare that there are no competing financial interests in carrying out this study.

5. References

[1] Yarden, Y Sliwowski, MX. Untangling the ErbB signalling network. *Nat Rev Mol Biol* 2:127-137, 2001.
[2] Pinkas-Kramarski, R Soussan, L Waterman, H Levkowitz, G Alroy, I *et al.* Diversification of Neu differentiation factor and epidermal growth factor signalling by combinatorial receptor interactions. *Embo J* 15: 2452-2467, 1996.
[3] Graus-Porta, D, Beerli, RR Daly, JM Hynes, NF. ErbB-2, the preferred heterodimerization partner on all Erb receptors, is mediator of lateral signalling. *Embo J* 16:1647-1655, 1997.
[4] Wang, LM Kuo, A Alimandi, M Veri, MC Lee, CC. *et al.* ErbB expression increases the spectrum and potency of ligand-mediated signal transduction through ErbB. *Proc Natl Acad Sci U S A* 95:6809-6814, 1998.
[5] Worthylake, R Opresko, LK Wiley, HS. ErbB-2 amplification inhibits down-regulation and induces constitutive activation of both ErbB and epidermal growth factor receptors. *J Biol Chem* 274:8865-8874, 1999.

[6] Desmedt, C Sperindle, J Piette, F Huang, W Jin, X. *et al*. Quantitation of HER2 expression or HER2:HER2 dimers and differential survival in a cohort of metastatic breast cancer patients carefully for trastuzumab treatment primarily by FISH. *Diagn Mol Pathol*. 18:22-29, 2009.

[7] Wolf-Yardin, A Kumer, N Zhang, Y Hautaniememi, S Zaman, M *et al*. Effects of HER2 overexpression on cell signalling networks governing proliferation and migration. *Mol Syst Biol* 2:55-60, 2006.

[8] Yuan, CX Lasut, AL Wynn, R Neff, NT Hollis, GF. Purification of Her-2 extracellular domain and identification of its cleavage site. *Protein Expr Purif.*29:217-22, 2003

[9] Molina, MA Codony-Servat, J Albanell, J Rojo, F Arribas, J. *et al*. Trastuzumab (herceptin), a humanized anti-Her2 receptor monoclonal antibody, inhibits basal and activated Her2 ectodomain cleavage in breast cancer cells. *Cancer Res*. 61: 4744-4749, 2001.

[10] Dankort D, Jeyabalan N, Jones N, Dumont DJ, Muller WJ. Multiple ErbB-2/Neu Phosphorylation Sites Mediate Transformation through Distinct Effector Proteins. J Biol Chem. Oct 19;276(42):38921-8, 2001

[11] Zhou, BP Liao, Y Xia, W Spohn, L Lee, MH. *et al*. Cytoplasmic localization of p21 Cip1/WAF1 by Akt-induced phosphorylation in HER-2/neu-over expressing cells. *Nature Cell Biol*; 3: 245-252, 2001.

[12] Fedi, P Pierce, JH di Fiore, PP. Efficient coupling with phosphotidylinositol 3-kinase, but not phospholipase C gamma or GTPase-activating protein, distinguishes ERB-3 signalling from that of other ERB/EGFR family receptors. *Mol Cell Biol* 14:492-500, 1994.

[13] Holbro T, Beerli RR, Maurer F, et al The ErbB2/ErbB3 heterodimer functions as an oncogenic unit: ErbB2 requires ErbB3 to drive breast cell proliferation. Proc Natl Acad Sci U S A 100:88933-8938 2003.

[14] Mohsin SK, Weiss HL, Gutierrez MC, Chamness GC, Schiff R, Digiovanna MP, Wang CX, Hilsenbeck SG, Osborne CK, Allred DC, Elledge R, Chang JC. Neoadjuvant trastuzumab induces apoptosis in primary breast cancer J Clin Oncolol 23: 7238-7240, 2005

[15] Kauraniemi P, Barlund M, Monni O, Kallioniemi. New amplified and Highly Expressed Genes Discovered in ERBB2 Amplicon in Breast Cancer by cDNA Microarrays Cancer Res: 61 8235-8240, 2001

[16] Mackay A, Jones C, Dexter T, LA Silva R, Bulmer K, JonesA, Simpson P, Harris RA, Jat PS, Neville AM, Ries LFL, Lakhani SR, O'Hare MJ c DNA microarray analysis of genes associated with ERBB2 (HER2/neu) overexpression in human mammary luminal epithelial cells. Oncogene 22; 2680-2688, 2003

[17] Kauraniemi P, Kallioniemi A. Activation of multiple cancer-associated genes at the ERBB2 amplicon in breast cancer.Endocr Relat Cancer. 13:39-49, 2006

[18] Arriola E, Marchio C, Tan DS, Drury SC, Lambros MB, Natrajan R, Rodriguez-Pinilla SM, Mackay A, Tamber N, Fenwick K, Jones C, Dowsett M, Ashworth A, Reis-Filho JS.

Genomic analysis of the HER2/TOP2A amplicon in breast cancer and breast cancer cell lines. Lab Invest. 88:491-503, 2008.

[19] Fornier M, Risio M, Van Poznak c, Seidman A Her2 testing and correlation with efficacy of trastuzumab therapy. Oncology 16: 1340-1348, 2002.

[20] Press MF, HER-2 gene amplification, HER-2 and epidermal growth factor mRNA and protein expression, and lapatinib efficacy in women with metastatic breast cancerClin Can Res. 14:7861-70 2008

[21] Chen QQ, Chen XY, Jiang YY, Liu J. Identification of novel nuclear localization signal within the ErbB-2 protein. Cell Res. 15:504-10, 2005.

[22] Wang C S, Lien HC, Xia W, Chen IF, Lo HW, Wang Z, Ali-Sayed M, Lee DF, Bartholomeusz G, Yang FO, Giri DK, Hung MC, Binding at transactivation of the COX-2 promoter by nuclear tyrosine kinase receptor Erb-2. Cancer Cell. 6: 251-261, 2004.

[23] Kumar R, Shepard HM, Mendelsohn J. Regulation of phosphorylation of the c-erbB-2/HER2 gene product by a monoclonal antibody and serum growth factor(s) in human mammary carcinoma cells. Mol Cell Biol. Feb;11(2):979-86, 1991.

[24] Lewis GD, Figari I, Fendly B, Wong WL, Carter P, Gorman C, Shepard HM. Differential responses of human tumor cell lines to anti-p185HER2 monoclonal antibodies. Cancer Immunol Immunother. Sep;37(4):255-63, 1993

[25] Carter P, Presta L, Gorman CM, Ridgway JB, Henner D, et al. Humanization of an ati-p185HER2 antibody for human cancer therapy. Proc Natl Acad Sci USA 89:4285-4289, 1992.

[26] Franklin, MC, Carey KD, Vajdos FF, Leahy DJ, de Vos AM, et al. Insights into ErbB signalling from the structure of the ErbB-pertuzumab complex. Cancer Cell 5:317-328, 2004.

[27] Agus DB, Gordon MS, Taylor C, et al. Phase I clinical study of pertuzumab, a novel HER dimerization inhibitor, in patients with advanced cancer. J Clin Oncol. 23:2534-2543, 2005.

[28] Freiss T, Scheuer W, Hasmann M. Superior antitumor activity after combination treatment with pertuzumab and trastuzumab against NSCLC and breast cancer xenograft tumors. Program and abstracts of the 31st European Society for Medical Oncology Congress; Istanbul, Turkey. September 29-October 3, 2006; Abstract 96PD.

[29] Chen WW, Input-output behaviour of ErbB signalling pathways as revealed by mass action model trained against dynamic data. Mol Syst Biol 5:239-58. 2006

[30] Xia W, Mullin RJ, Keith BR, Liu LH, Ma H, et al. Anti-tumor activity of GW572016: a dual tyrosine kinase inhibitor blocks EGF activation of EGFR/erbB2 and downstream Erk1/2 and AKT pathways. Oncogene. Sep 12;21(41):6255-63, 2002.

[31] Brockhoff G, Heckel B, Schmidt-Brueken E, Plander M, Hofstaedter F. Differential impact of Cetuximab, Pertuzumab and Trastuzumab on BT474 and SK-BR-3 breast cancer cell proliferation. Cell Prolif 40:488-507, 2007.

[32] Scaltriti M Lapatinib, A HER2 tyrosine kinase inhibitor, induced stabilization of HER2 and potentiates trastuzunmab-depedent cell cytotoxicity. Oncogene. 28: 803-14 2009

[33] Welle S, Welt S, Shay T, Lanning C, Horton K, et al. The use of gene array analysis to determine down-stream molecular expression patterns of trastuzumab (T) treatment. J Clin Onc, 2007 ASCO Annual Meeting Proceedings Part I, 25:14135,2007

[34] Mosesson Y, Mills GB, Yarden Y. Derailed endocytosis: an emerging feature of cancer. Nat Rev Cancer 9: 835-850, 2009.

[35] Klapper LN, Waterman H, Sela M, Yarden Y. Tumor-inhibitory antibodies to HER-2/ErbB-2 may act by recruiting c-Cbl and enhancing ubiquitination of HER-2. Cancer Res 60:3384-8, 2000.

[36] Binhua PZ, Yong L, Weiya X, Yiyu Z, Spohn B, et al. Her-2/ neu induces p53 ubiquination via Akt-mediated MDM2 phosphorylation. Nat Cell Biol. 3:973-982, 2001.

[37] Vogel CL, Cobleigh MA, Tripathy D, Gutheil JC, Harris LN, et al. First-line Herceptin monotherapy in metastatic breast cancer. Oncology. 61 Suppl 2:37-42, 2001.

[38] Moasser MM, Basso A, Averbuch SD, Rosen N. The tyrosine kinase inhibitor ZD1839 ("Iressa") inhibits HER-2 driven signaling and suppresses the growth of HER2-overexpressing tumor cells. Cancer Res. 61:7184-8, 2001.

[39] Moulder SL, Yakes FM, Muthuswamy SK, Bianco R, Simpson JF. Epidermal growth factor (HER1) tyrosine kinase inhibitor ZD1839 (Iressa) inhibits HER/neu (erbB2)-overexpressing breast cancer cells in vitro. Cancer Res. 2001. 61:8887-95.

[40] Sergina NV, Rausch M, Wang D, Blair J, Hann B, et al. Escape from Her-family tyrosine inhibitor therapy by kinase inhibitor therapy by the kinase-inactive HER3. Nature 445:437-441 2007.

[41] Wong KK, Fracasso PM, Bukowski RM, Lynch TJ, Munster PN, et al. A phase I study with neratinib (HKI-272), an irreversible pan ErbB receptor tyrosine kinase inhibitor, in patients with solid tumors. Clin Cancer Res. Apr 1;15(7):2552-8, 2009.

[42] Karaman MW, Herrgard S, Treiber DK, Gallant P, Atteridge CE, et al. A quantitative analysis of kinase inhibitor selectivity. Nat Biotechnol. Jan;26(1):127-32, 2008.

[43] Schulze WX, Deng L, Mann M. Phosohotyrosine interactome of the erbB-receptor kinase family, Mol Syst Biol. 10:1-13, 2005.

[44] Strohecker AM, Yehiely F, Chen F, Cryns VL. Caspase cleavage of HER-2 releases a Bad-like cell death effector. J Biol Chem. Jun 27;283(26):18269-82, 2008.

[45] Pedersen K A naturally occurring HER2 carboxyl-terminal fragment promotes mammary tumor growth and metastases. Mol Cell Biol 29:3319-31 2009

[46] Clynes RA, Towers TL, Presta LG, Ravetch JV. Inhibitory Fc receptors modulate in vivo cytoxicity against tumor targets. Nat Med. Apr;6(4):373-4, 2000.

[47] Gennari R, Menard S, Fagnoni F, Ponchio L, Scelsi M, et al. Pilot study of the mechanism of action of preoperative trastuzumab in patients with primary operable breast tumors overexpressing HER2. Clin Cancer Res. Sep 1;10(17):5650-5, 2004.

[48] Segerling M, Ohanian SH, Boros T. Chemotherapeutic drugs increase killing of tumor cells by antibody and complement. Science 188: 55-57, 1975.

Protein Kinases and Phosphatases in Cell Cycle Regulation

Phosphorylation Mediated Regulation of Cdc25 Activity, Localization and Stability

C. Frazer and P.G. Young

Additional information is available at the end of the chapter

1. Introduction

Dual specificity phosphatases of the Cdc25 family are critically important regulators of the cell cycle. They activate cyclin-dependent kinases (CDKs) at key cell cycle transitions such as the initiation of DNA synthesis and mitosis. They also represent key points of regulation for pathways monitoring DNA integrity, DNA replication, growth factor signaling and extracellular stress. Since their mis-regulation allows cells to function in a genetically unstable state, it is not surprising that these phosphatases are involved in transformation to a cancerous state. Cdc25 phosphatases are heavily regulated by phosphorylation. Many regulatory phosphorylation sites on Cdc25 influence catalytic activity, substrate specificity, subcellular localization and stability. This chapter summarizes the current literature on the phospho-regulation of these proteins.

2. Yeast genetics and Xenopus oocyte maturation – Setting the stage

The study of cell division in eukaryotes was dramatically changed with the isolation of temperature sensitive "cell division cycle" (*cdc*) mutants of the yeasts *Schizosaccharomyces pombe* and *Saccharomyces cerevisiae*. These mutants arrested uniformly at a particular cell cycle stage and uncoupled cell growth from cell cycle progression.[1-3] *S. pombe* cells are cylindrical cells growing from the tips while keeping a constant diameter.[3] At the restrictive temperature, fission yeast *cdc* mutants arrest at their restriction point and become abnormally elongated. Within the fission yeast *cdc* collection was the *cdc25-22* mutation which arrests at mitotic entry when incubated at its restrictive temperature.[3] The collection also included *cdc9-50* which divided at approximately half the length of a wildtype cell; it was later renamed *wee1-50* in reference to its small size.[4] The *wee1-50* mutation suppresses the cell cycle phenotype of *cdc25-22* demonstrating that these two proteins act in opposition during the G2/M transition.[5] Analysis of these mutations showed that Cdc2 is rate limiting for mitotic entry and is inhibited by Wee1 and activated by Cdc25.[6,7] Regulation of Cdc2 by Wee1 and Cdc25 in fission yeast was one of the first connected pathways in cell cycle research in any organism. The gene names Cdc25 and Wee1 are used in almost all organisms today.

Concurrently, Maturation Promoting Factor (MPF) was discovered through a series of elegant cytoplasmic transfer experiments conducted on frog, starfish and sea urchin oocytes [8-10]. Immature *Xenopus* oocytes arrest at prophase I. Progesterone induced MPF activation induces maturation whereby oocytes complete meiosis I and arrest at meiotic metaphase II.[9,11] Microinjection of cytoplasm from a mature oocyte into an immature oocyte also induces maturation.[8,9] This bioassay for maturation promoting factor (MPF) activity allowed it to be tracked through the meiotic and later mitotic cycles. Activity falls after anaphase I, rising again as cells enter prophase II. Although MPF activity is high, mature oocytes arrest due to the presence of a second soluble factor called Cyto-Static Factor (CSF).[9] Subsequent entry of the sperm nucleus causes an influx of extracellular calcium, ultimately removing CSF inhibition and serving as the trigger for completion of meiosis and the start of mitotic divisions. Extracts from prophase II arrested mature oocytes can be induced to undergo alternating DNA synthesis and mitotic phases by the addition of calcium, providing an excellent cell free system for studying MPF regulation.[12] A peak of MPF activity accompanies each round of mitosis, reaching maximal activity during meiotic prophase, followed by a catastrophic drop as chromosomes segregate at anaphase. Fertilization is followed by a rapid succession of 12 rounds of DNA synthesis and cell division with no intervening gap phases, driven solely by maternally derived mRNA.[13] Gap phases are re-established at the mid-blastula transition followed by typical somatic cell cycles.[14] MPF consists of three proteins: Cdk1 kinase, cyclin B (Cdc2 [19-21] and Cdc13 [22,23] respectively in fission yeast) and the small regulatory protein Suc1.[15-19] The activator of both MPF in *Xenopus* and the Cdc2-Cdc13 complex in *S. pombe* is Cdc25 phosphatase.[24,25]

The cytoplasmic transfer experiments showed that MPF activity correlates with increased protein phosphorylation in donor oocytes [26], targeting a large number of nuclear proteins.[27,28] MPF is a histone H1 kinase, an activity which is still used today to measure CDK function.[29] MPF induces many of the cytological changes associated with mitosis such as nuclear envelope breakdown, chromosome condensation and mitotic spindle formation.[30-32] Mass spectrometry has shown that the single Cdk1 homologue in budding yeast (CDC28) phosphorylates 547 sites on 308 proteins *in vivo*.[33] The regulation of only about 75 of these CDC28 substrates has been examined in any detail.[34] Undoubtedly, the network of CDK mediated phosphorylation events in vertebrate cells will turn out to be far more complex.

3. Taxonomic distribution, duplication and divergence of Cdc25 homologues

Cdc25 is present in all eukaryotic cells. The yeasts, *S. cerevisiae* and *S. pombe*, possess a single Cdc25 protein (referred to as MIH1 in the former).[35] Duplication and divergence led to three Cdc25 paralogues in vertebrates, Cdc25A, Cdc25B, and Cdc25C.[36,36-40] Cdc25A acts early in the cell cycle, regulating the G1/S transition, whereas Cdc25B and Cdc25C act at G2/M. Human Cdc25A, Cdc25B and Cdc25C have several isoforms through alternative splicing.[41-44] Cdc25C was discovered based on sequence similarity to fission yeast Cdc25 by using degenerate PCR primers to conserved residues in the catalytic region.[38] Human Cdc25A and Cdc25B were cloned by complementation of the temperature sensitive *cdc25-22^ts* allele of *S. pombe*, demonstrating the strong conservation of function in this protein family.[37] *Xenopus* Cdc25A and Cdc25C were initially discovered based on sequence similarity to human Cdc25C.[40,45] Injection of recombinant *Xenopus* Cdc25A or Cdc25C into oocytes induces their maturation, the first direct indication that this protein was a positive regulator of MPF.[40,45,46] *Xenopus* Cdc25C also complements the *cdc25-22^ts* mutation in fission yeast.[45] Four Cdc25 homologues are present in *Caenorhabditis* [47], whereas *Drosophila* contains two, *string* and *twine*.[48,49] The conserved C-terminal domain of *string* contains the intrinsic phosphatase activity against a Cdk1-derived peptide containing phosphorylated Y15 and against the synthetic substrate *p*-nitrophenyl phosphate.[50] Few plant Cdc25 homologues can be found by sequence comparison in the NCBI database and those only in unicellular algae, eg *Ostreococcus tauri*.[51] In *Arabidopsis thaliana* [52] the full length protein is only 146 residues long representing only the phosphatase domain and it shows marginal similarity to vertebrate and fission yeast Cdc25. Overexpression of *A. thaliana* Cdc25 causes premature mitotic entry in fission yeast.[53]

4. Regulation of cell cycle transitions by Cdc25

4.1. The fission yeast G2/M transition – The prototype Cdc25/cyclin-CDK circuit

Regulation of the transition from G2 to mitosis in *S. pombe* is typical of the positive feedback loops seen between Cdc25 and CDKs in all systems. Cdc2 (Cdk1) drives all cell cycle transitions in fission yeast.[20] During G2 and early mitosis Cdc2 complexes with the B-type cyclin Cdc13 [54,55], the only essential cyclin in fission yeast.[56] Whereas Cdc2 is constitutively expressed throughout the cell cycle [57], Cdc25 and Cdc13 begin to accumulate in G2 and are degraded during mitotic entry.[58-60] Prior to M-phase, the Cdc2-Cdc13 complex is kept inactive by phosphorylation of threonine 14 (T14) and tyrosine 15 (Y15).[61,62] Y15 is phosphorylated by the S-phase specific kinase Mik1 [63,64] and T14 and Y15 are modified by Wee1.[63,65-67] In G2 Cdc25 translocates to the nucleus via the importin-β homologue Sal3.[68] After cells grow to a critical size for mitotic entry, Cdc25 becomes active and dephosphorylates Cdc2 Y15.[69] Cdc25 is activated and hyperphosphorylated by Cdc2 as the cell enters mitosis.[58, 70] CDKs phosphorylate a serine or threonine residue in the context of a consensus site (S/T)PX(K/R).[71,72] The positions of the Cdc2 phosphorylation sites on Cdc25 have not been determined, but mutagenesis of 15 potential CDK consensus sites abrogates phosphorylation by Cdc2-Cdc13.[73] In *S. pombe*, the feedback loop between Cdc25 and Cdc2-Cdc13 is strengthened by the involvement of the *S. pombe* Polo kinase homologue Plo1 which phosphorylates Cdc25 downstream of Cdc2 activation.[74] After the metaphase-anaphase transition, Polo kinase activation encourages cyclin degradation through activation of the anaphase promoting complex (APC).[75] Thus, this kinase simultaneously ensures that Cdc2 activation will be robust, and brief, as its cyclin partner Cdc13 is degraded by the APC immediately following anaphase initiation.

4.2 Vertebrate cell cycle

While fission yeast Cdc25 is solely involved in the G2/M transition, Cdc25 orthologues in vertebrates also play a role in G1 and S-phase progression. Vertebrates have several CDK-cyclin complexes which participate in these transitions through positive feedback loops with Cdc25. In addition to associating with cyclin B, Cdk1 associates with cyclin A during late S-phase and G2.[76] A second CDK, Cdk2, was discovered as a cDNA which could complement the loss of the budding yeast Cdk1 homologue, CDC28.[77] Cdk2 functions early in the cell cycle and is likewise negatively regulated by phosphorylation of Y15.[78,79] It forms a complex with Cyclin A and Cyclin E.[80,81] Cdk4 and Cdk6 operate early in G1 in association with D-type cyclins.[82] In many cases a particular cyclin class (ie. A, B, D, E) has multiple members. For the sake of clarity, cyclins will be referred to by their subtype only. Furthermore, only the CDKs and cyclins directly responsible for cell cycle transitions in concert with Cdc25 orthologues will be discussed. For instance, Cdk1/cyclin B is activated by a CDK-Activating Kinase (CAK), a complex of Cdk7 and cyclin H [83], but as CAKs are not activated by Cdc25 they are outside the scope of this review.

4.2.1. Cell cycle re-entry from G_0

Most somatic cells spend their time in G_0. Cells in G_0 may commit to entry into the cell cycle when they receive stimuli in the form of growth factors The cells then deactivate the cell cycle repressors which have kept them in G_0 and transcribe the positive regulators of the next cell cycle transition, G1/S.

In non-dividing cells, the Retinoblastoma protein (Rb) binds to E2F thus preventing transcription of genes required for cell cycle progression including cyclin D, cyclin A and Cdc25A.[84-87] (Figure 1) After exposure to growth factors, cells in G_0 re-enter the cell cycle through activation of the Ras pathway via the Raf/MAP (Mitogen Activated Protein) kinase pathway.[88] This leads to the degradation of the Cdk4 inhibitors p15^{INK4B} and p16^{INK4A} and the Cdk2 inhibitors p27^{KIP}, p21^{CIP} and induction of cyclin D. Cdk4 and Cdk6 bound to cyclin D inhibit the Rb protein [89] allowing transcription of cyclin E and cyclin A.[89,90] Cdc25A activates Cdk4-cyclin D but not Cdk6-cyclin D *in vitro*.[91] Cyclin E associates with Cdk2 in late G1 and helps complete the inhibition of Rb.[89,92]

4.2.2. The G1/S transition

Cdc25A is transcribed following relief of Rb-mediated transcriptional repression, reaching its maximal level at the end of G1 and dephosphorylating Y15 on Cdk2.[87,91] Cdk2 is phosphorylated on Y15 by Wee1.[93] Dephosphorylation of Cdk2 takes place via both

Cdc25A and Cdc25B.[94-97] Cdk2 immunoprecipitated from cell lysates where Cdc25A has been overexpressed has high histone H1 kinase activity and low levels of Y15 phosphorylation.[91] Such overexpression accelerates entry into S-phase through activation of Cdk2-cyclin E.[91,98] DNA-synthesis can be blocked in these cells by injecting them with anti-Cdc25A antibodies.[99] Cdk2-cyclin E and Cdc25A are mutually activated by a positive feedback loop allowing passage of the G1/S boundary.[95] Depletion of Cdk2 or cyclin E prevents phosphorylation of Cdc25A and recombinant Cdc25A can be activated by Cdk2-cyclin E *in vitro*.[95] In addition phosphorylation of Cdc25A by Cdk2 destabilizes the phosphatase. Conversely, exposure of cells to CDK inhibitors roscovitine and olomoucine causes stabilization of Cdc25A.[99]

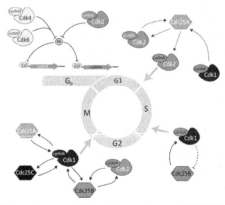

Figure 1. Positive feedback loops between Cdk-cyclin complexes and Cdc25 family members drive cell cycle transitions.

S-phase initiation requires activation of DNA replication proteins by Cdk2-cyclin A. Injecting G1 cells with anti-cyclin A antibodies stops entry into S-phase.[100] Phosphorylation of the essential DNA replication initiator Cdc6 by Cdk2-cyclin A leads to its nuclear import.[101,102] Cdk2 is recruited to chromatin by the replication initiation factor Cdc45 where it phosphorylates histone H1 and induces chromosome de-condensation.[103] As S-phase progresses high Cdk2-cyclin A activity induces degradation of Cdc6, preventing re-initiation of DNA synthesis at origins which have already fired.[102]

Cdc25B has a role late in S-phase as Cdc25B immunoprecipitated from late S-phase HeLa cell extracts is phosphorylated, activated, and able to dephosphorylate Cdk2-cyclin A.[97] The murine homologue of Cdc25B purified from S-phase extracts promotes cyclin A and cyclin E associated histone H1 kinase activity *in vitro*.[94] In addition, siRNA knockdown of human Cdc25B causes a delay in initiation of DNA synthesis.[104] The Cdk1-cyclin A complex regulates late origin firing in S-phase. A constitutively activated Cdk1 allele increases firing of late S-phase origins, whereas a loss of function temperature sensitive allele of Cdk1 has a defect in late origin firing at the restrictive temperature.[76]

4.2.3. The G2/M transition

Unlike the simple circuit of Cdc25-Cdc2 activation in fission yeast, the vertebrate G2/M transition involves a series of interconnected loops with positive and negative inputs from a variety of pathways. This led to what could be considered the "traditional model" of G2/M transition in vertebrates with respect to Cdc25 regulation by CDK-cyclin complexes. In reality things may be more complex. Cdc25C is not explicitly required for mitotic entry; siRNA knockdown of Cdc25C does not prevent the G2/M transition.[105] In addition, mouse lines which lack Cdc25B and/or Cdc25C are viable with the only obvious phenotype being a defect in oocyte maturation observed in $cdc25B^{-/-}$ mice.[106-108] In addition, the creation of CDK and cyclin knockout mice has revealed a network of compensatory mechanisms which cloud the traditional model. $Cdk2^{-/-}$, $Cdk4^{-/-}$, or $Cdk6^{-/-}$ mice show some abnormalities as adults, but are not embryonic lethal indicating that these Cdks are not individually essential for cell division.[92] There is considerable redundancy. Mitotic entry is best viewed as a three step process. First, Cdc25B and CDK-cyclin A are activated followed by basal activation of Cdk1-cyclin B at the centrosome.[109] Second, Cdk1-cyclinB, Cdc25B and Cdc25C localize to the nucleus. Third, Cdk1-cyclin B and Cdc25C mutually activate. Cdk1-cyclin B then induces overt mitotic events such as breakdown of the nuclear envelope and spindle formation. [81,110,111]

4.2.4. Activation of Cdc25B and Cdk1/2-cyclinA

Phosphorylation of human Cdc25B by Cdk1-cyclin A during G2 causes Cdc25B activation but also destabilizes the protein.[112,113] Cdc25B activates Cdk2-cyclinA in a positive feedback loop.[114] Cdk2-cyclin A mediated destabilization of Cdc25B has not been reported, although the Cdk1 and Cdk2 kinase complexes modify Cdc25B to approximately the same degree and have a similar set of substrates.[113,115] However, Cdk2 is the preferred binding partner of cyclin A and is more active than Cdk1-cyclin A during G2.[80,100] Cdk2 cyclin A has two peaks of activation, one during S-phase and one prior to G2/M. [80,116] Inhibition of Cdk2-cyclin A delays mitotic entry.[109] Depletion of Cdk1-cyclin A, activation of the CDK inhibitor p21^{cip} , or addition of an inhibitory ATP analogue destabilizes Cdc25B in $Sf9$ cell extracts.[112] Similarly, in cycloheximide treated cells Cdc25B, but not Cdc25A or Cdc25C, is unstable and uniquely labile during the cell cycle.[117]

Cyclin B accumulates throughout G2 but Cdc25A and Cdc25B are required to induce formation of the Cdk1-cyclin B complex at G2/M.[118] Cdk1-cyclin B interaction occurs earlier in G2 when Cdc25A or Cdc25B are overexpressed. Cdk1 and cyclin B are almost exclusively cytoplasmic during interphase with a small portion of the complex associating with the centrosome at G2/M.[119-121] A population of Cdk2-cyclinA is likewise localized to the centrosomes prior to prophase.[116] In $Xenopus$, centrosomal localization of Cdk1-cyclin B requires Aurora kinase.[122] Aurora is involved in a diverse set of mitotic events such as spindle assembly, centrosome maturation, and cytokinesis.[123] Aurora can also phosphorylate Cdc25B at S353 $in vivo$, activating the phosphatase.[124] (Table 1) Although some Cdk1-cyclin B is activated by Cdc25B [120,125], this does not represent a full mitotic activation and is only sufficient for progression as far as prophase. Injection of a dominant-negative mutant of Cdc25B into HeLa cells causes arrest in prophase with condensed chromosomes and disassembled nucleoli, but without nuclear envelope breakdown [109]. Microinjection of Cdk1-cyclin B lets cells pass this block and complete mitosis.

Species	Member	Kinase	Site(s)	References
S. pombe	Cdc25	Cds1/Chk1/Srk1	S99, S148, S178, S192, S204, S206, T226, S234, S359, T561, S567, T569	[60, 156, 157, 162]
Human	Cdc25A	Cdk1-cyclin B	S18, S40, S88, S116, S261, S283, S321	[149, 151, 167]
		Chk1	S76, S124, S178, T507	[168-173, 200]
		Chk2	S124, S278	[173, 175]
		Casein Kinase 1ε	S82	[199]
		Casein Kinase 1α	S79, S82	[198]
		p38	S76, S124	[168, 175]
		GSK-3β	S76	[193, 194]
		NEK11	S82, S88	[187]
		Plk3	S80	[193, 194]
	Cdc25B*	Cdk1-cyclinB	S50, S160, S321	[135, 136, 246]
		Chk1	S151, S230, S323, S563	[236, 238, 239]
		Aurora	S353	[124]
		Casein Kinase 2	S186, S187	[235]
		JNK	S101, S103	[280]
		MEK/ERK	S249	[259]
		p38	S323	[208]
		MK2	S323	[210]
		Plk1	T167, S209, T404, S465	[136]
	Cdc25C	Cdk1/cyclin B	T48, T67, S122, T130, S168, S214	[42, 143, 145, 246]
		Chk1	S216, S247, S263	[219-221, 225, 226]
		Chk2	S216	[209, 242]
		Casein Kinase 2	T236	[234]
		MK2	S216	[210]
		C-TAK1	S216	[276]
		JNK	S168	[281, 282]
		MEK/ERK	S216	[278]
		Plk1	S198	[140]
		Plk3	S191, S198	[141]
Xenopus	Cdc25A	Chk1	S73, T504	[201, 207]
	Cdc25C	Cdk1/cylinB Cdk1-cyclinA	T48, T67, T138, S205, S285;	[193, 247, 249]
		Chk1	S287, T533	[207, 229-231, 241]
		p42	S48, T138, S205	[272]
		p90^{rsk}	S287	[269, 270]
		Rsk2	S317, S318, S319	[271]

Table 1. Summary of known phosphorylation sites on Cdc25 family members and the kinases responsible

4.2.5. Everybody into the nucleus

Cyclin B has a cytoplasmic retention sequence which is sufficient to induce cytoplasmic localization of the normally nuclear protein and contains a nuclear export signal (NES).[126]

Nuclear export is blocked by phosphorylation of S126 at the end of prophase.[127-129] Cyclin B is phosphorylated by Cdk1-Cyclin B as starfish oocytes pass the prophase II to metaphase II arrest.[130] Human cyclin B S126 is followed by a proline residue suggesting Cdk1-cyclin B autophosphorylation. Cyclin B is phosphorylated by the *Xenopus* Polo kinase homologue Plx1 on S133 and S147 (homologous to human cyclin B S177 and S181) enhancing its nuclear import.[125,131] Initial Cdc25B mediated activation of Cdk1-cyclin B may result in a priming autophosphorylation of S126 followed by docking of Polo kinase and inhibition of nuclear export. Such a Cdk1 mediated priming phosphorylation has been shown to induce Plx1 phosphorylation of Cdc25B (see below). Human Polo kinase Plk1 also deactivates Cdk1-cyclin B inhibitory kinases. Plk1 inhibits the cytoplasmic Cdk1 inhibitory kinase Myt1.[132,133] Wee1 is phosphorylated on S53 and S123 by Plk1 and Cdk1-cyclin B, respectively, resulting in its degradation by β-TrCP.[134]

The activated cytoplasmic pool of Cdk1-cyclin B phosphorylates Cdc25B on S160 inducing its nuclear import. [135] (All phosphorylated residues on human Cdc25B are numbered according to the sequence of the longest splice variant Cdc25B3). Overexpression of human Cdc25B causes an increase in cells with condensed chromatin, whereas overexpression of Cdc25B-S160G does not induce mitotic entry. S160 phosphorylation does not affect the *in vitro* activity of human Cdc25B against a fluorescein diphosphate substrate, but instead positively regulates its nuclear import. *In vitro* phosphorylation assays show Cdk1-cyclin B can phosphorylate residues on Cdc25B which are not targeted by Cdk1-cyclin A or Cdk2-cyclin A.[113] Interestingly, Cdk1-cyclin A cannot phosphorylate Cdc25B that has previously been phosphorylated by Cdk1-cyclin B.[112]

Plk1 is involved in human Cdc25B nuclear import following the initial activation of Cdk1-cyclin B. After addition of the Plk1 inhibitor thiophene benzimidazole, nuclear accumulation of GFP-Cdc25B is reduced. Conversely, expression of a constitutively active Plk1 mutant enhances Cdc25 nuclear localization.[136] Co-overexpression of Plk1 and Cdc25B in U2OS osteosarcoma cells induces chromosome condensation to a greater degree compared to cells expressing Plk1 or Cdc25B alone. This is partially dependent on the presence of a functional nuclear localization signal (NLS) in Cdc25B. Plk1 docking to Cdc25B requires prior phosphorylation of S50 by Cdk1/cyclin B.[137] Mass spectrometry identified thirteen phosphorylated Plk1 sites on Cdc25B *in vitro* and showed that T167, S209, T404, S465 and S513 appear to be particularly strong targets.[137] Plk1 is required for Cdk1-cyclin B activation, at least in part by negatively regulating Cdc25C nuclear export. Depletion of Plx1 from oocyte extracts prevents the activation of Cdc25C and Cdk1-cyclin B.[138] Unlike Cdc25B, phosphorylation of Cdc25C by Cdk1 is not a prerequisite for Plk1 targeting the phosphatase; Plk1 affects Cdc25C phosphatase activity and its localization. *In vitro* phosphorylation of Cdc25C enhances its ability to dephosphorylate kinase dead Cdc2 on Y15.[139] Plk1 phosphorylates human Cdc25C on S198 which resides within the nuclear export signal (NES) of the phosphatase, promoting its nuclear localization.[140] Polo kinase family member Plk3 also interacts with human Cdc25C and phosphorylates it on S191, and S198 to a lesser degree.[141] Substitutions S191D, S198D or Plk3 overexpression result in constitutive nuclear localization, while siRNA knockdown of Plk3 or substitutions S191A and S198A leads to nuclear exclusion.[141] Another polo family member, Plk4, phosphorylates human Cdc25C on undetermined sites.[142]

4.2.6. Full activation of Cdk1-CyclinB and Cdc25C

Activated Cdk1-cyclin B phosphorylates human Cdc25C on T48, T67, S122, T130, S168 and S214 *in vitro* and *in vivo*, driving a positive feedback loop which culminates in the phosphorylation of mitotic Cdk1 substrates.[42,143] Cdk1 phosphorylated Cdc25C has an increased Cdk1 Y15 phosphatase activity.[143] Recombinant Cdc25C can activate Cdk1 thereby increasing its histone H1 kinase activity.[144] Use of phospho-specific antibodies recently showed that phosphorylation of T48, T67 and T130 occur on spatially separate pools of human Cdc25C.[145] T67-phosphorylated Cdc25C is chromatin associated from prophase until telophase while T130-phosphorylated Cdc25C localizes to the centrosomes. T130 phosphorylation creates a Plk1 binding site on Cdc25C. Three distinct pools of T48, T67 and T130 mono-phosphorylated Cdc25C protein can be detected by immunoprecipitation with each individual phospho-specific antibody; Cdc25C pulled down with one antibody is not bound by the other two. In *Xenopus* oocyte extracts Cdc25C is heavily phosphorylated while cells undergo germinal vesicle breakdown (meiosis I).[45] *Xenopus* Cdc25C residues T48, T67, T138, S205 and S285 are phosphorylated by Cdk1-cyclin A and Cdk1-cyclin B *in vitro*.[146] Mutation of the major targets, T48, T67 and T138, to alanine prevents *Xenopus* Cdc25C activation *in vitro*.[147] Although G2/M is not normally associated with Cdc25A function, it also contributes to Cdk1-cyclin B activation. Depleting cells of Cdc25A reduces Cdk1-cyclin B activation by approximately fifty percent.[148] In cells arrested in mitosis by addition of nocodazole, a microtubule polymerization inhibitor and spindle poison, Cdc25A is phosphorylated by Cdk1-cyclin B on S18 and S116. Substitution of these residues with alanine leads to Cdc25A instability. Cdc25A S40, S88, S261 and S283 are also Cdk1-cyclin B phosphorylated *in vitro*.[149]

4.2.7. Mitotic exit

Following chromosome alignment on the metaphase plate a cascade of APC mediated degradation events occurs to reset conditions for the start of the next cell cycle.[150] The APC regulates two important processes required for completion of mitosis. First, it targets Securin, the inhibitory subunit of Separase, which is responsible for Cohesin cleavage and chromosome separation. Second, it targets cyclin A and cyclin B for destruction, inactivating Cdk1. Cyclin B is degraded after the metaphase-anaphase transition while cyclin A is degraded during metaphase. [121] Cdk1 mediated phosphorylation of Cdc25 paralogues is reversed by Cdc14 family phosphatases. Cdk1-cyclin B phosphorylates human Cdc25A S18, S40, S88, S116, S261 and S283 *in vitro*.[149] Of these sites, human Cdc14A dephosphorylates S116, Cdc14B targets S88 and S261, while both phosphatases can dephosphorylate S40. Cdc14A was independently identified as also dephosphorylating Cdc25A S321.[151] siRNA knockdown of Cdc14B leads to accumulation of phosphorylated Cdc25B and Cdc25C.[149] In *S. pombe* Cdc14 homologue Clp1 dephosphorylates Cdc25 and this is required for its ubiquitination and APC mediated proteolysis at the end of mitosis.[73] Similarly, in vertebrates Cdc25A and Cdc25B are targeted by the APC at mitotic exit.[152,153]

5. Cdc25 phosphorylation by the DNA damage and replication checkpoint

DNA damage causes activation of checkpoints which delay cell cycle transitions to allow sufficient time for repair. Stalling of replication forks causes a similar cell cycle arrest, with additional need for stabilizing replication forks and/or modulating replication origin firing until the cause of the stalling is eliminated. Checkpoint effector kinases impinge on the central cell cycle machinery by phosphorylating Cdc25. This modification variously inhibits Cdc25 phosphatase activity, induces degradation or creates binding sites for 14-3-3 proteins which modify localization of the protein.

5.1. Fission yeast

Cell cycle arrest following DNA damage requires that Cdc2 is kept in a Y15-phosphorylated, inhibited state.[154] Cdc25 is inhibited through phosphorylation by Chk1 and Cds1 kinases in response to DNA damage and replication fork arrest, respectively.[155-157] Cells over-expressing Cdc25 or expressing a Y15F phospho-mimetic mutation of Cdc2 fail to arrest cell cycle progression after exposure to ionizing radiation.[154] In the absence of Cds1, Chk1 can cause cell cycle arrest following stalling of replication forks by hydroxyurea (HU) exposure. Cells lacking both kinases are unable to arrest.[155] Cds1 also phosphorylates multiple substrates to stabilize stalled replication forks and prevent the occurrence of inappropriate recombination events.[158] Upstream regulation of the DNA damage and replication checkpoint pathway occurs through activation of the ATM (Ataxia-telangiectasia mutated) homologue Rad3 through a well conserved signaling cascade.[159]

Phosphorylation of Cdc25 by Chk1 and Cds1 creates binding sites for the 14-3-3 homologues Rad24 and Rad25.[160,161] Phosphorylation and 14-3-3 binding stabilizes Cdc25, a phenomena referred to as "stockpiling", thought to allow the cell to rapidly re-enter the cell cycle once replication or DNA damage arrest has been lifted.[70] The first Chk1/Cds1 phosphorylated Cdc25 residue identified, S99, partially impairs the replication and DNA damage checkpoint when mutated to alanine.[157] S99 modified Cdc25 is also phosphorylated on S192 and S359 by Cds1 and Chk1 *in vivo* and *in vitro*.[156] By phosphorylating Cdc25 *in vitro* with Cds1, nine additional sites were identified by mass spectrometry (S148, S178, S204, S206, T226, S234, T561, S567, T569) as well as the three sites previously known.[162] Nine sites between S99 and S359 are distributed through the poorly conserved N-terminal two thirds of the protein, while three sites reside in the extreme C-terminus. Alanine substitutions of all nine sites in the amino two thirds of Cdc25, *cdc25(9A)*, overrides the DNA replication checkpoint when expressed under control of a relatively weak heterologous promoter.[162] However, when *cdc25(9A)* is expressed from the native *cdc25+* locus under the control of its own promoter it does not cause a cell cycle phenotype, and the cell is checkpoint competent.[60] In addition, Cdc25(9A) is unstable following replication arrest suggesting redundancy in checkpoint control, such that Cdc25 which cannot be inhibited by Cds1 is eliminated from the cell. The S-phase specific Cdc2-Y15 kinase Mik1 is sufficient to prevent mitotic entry in these cells. Alanine substitutions of only the three C-terminal Cdc25 sites (T561, S567, T569) have a clear replication checkpoint defect in a *mik1*⁻ background and appear to be involved in maintenance, but not establishment, of the DNA damage checkpoint. [305]

5.2. Vertebrate Cdc25 regulation by DNA damage and replication checkpoints

In vertebrate cells detection of DNA damage is relayed through ATM and ATR (ATM-Related) to two checkpoint effectors, Chk1 and Chk2. While the *S. pombe* ATM homologue Rad3 is involved in activation of both Chk1 and Cds1; DNA damage signaling in vertebrates shows separation of ATR-Chk1 and ATM-Chk2 axes.[163,164] The target of these effector proteins is determined by the cell cycle stage at which the damage occurs and the nature of the damage itself. ATM-Chk2 signaling is initiated by double strand breaks, while ATR-Chk1 is activated by stalled replication forks and single stranded breaks. The p38 MAP kinase pathway is critical for cell cycle arrest following UV induced DNA damage. (Figure 2)

5.2.1. G1/S and Intra-S checkpoints

The G1/S checkpoint prevents the start of DNA synthesis in the presence of DNA damage while the Intra-S checkpoint protects replication forks, prevents activation of late replication origins, and keeps the cell from entering mitosis until S-phase is completed. G1-S checkpoint arrest is manifested through inhibition of Cdc25A, thus preventing activation of Cdk4-cyclin D and Cdk2-cyclin E. The checkpoint also activates p53 resulting in the induction of the Cdk2 inhibitor p21^{CIP} and the targeting of Cyclin E to the SCF complex to reinforce Cdk2 inhibition.[165] In rat fibroblasts UV induced DNA damage during G1 results in cell cycle arrest at the G1/S transition requiring inhibition of Cdc25A and phosphorylation of Cdk4-Y14.[166] In U2OS osteosarcoma cells Cdk2-cyclin E kinase activity decreases and Y15 phosphorylation increases coincident with Cdc25A degradation following UV exposure.[167] Conversely, Cdc25A overexpression in UV exposed U2OS osteosarcoma cells results in bypass of the checkpoint and dephosphorylation of Cdk2-cyclin E. Cdc25A inhibition involves a combination of destabilization and inhibition of phosphatase activity by Chk1.[167] Treatment with caffeine (an ATM/ATR inhibitor) or the Chk1 inhibitor UNC-01, or depletion of Chk1, stabilizes the phosphatase.[167,168] In humans, Chk1 phosphorylates Cdc25A S76, S124 , S178, and T507.[168-170] Cdc25A catalytic activity is reduced three-fold when it is phosphorylated by hChk1 *in vitro*.[169] S76 and S124 are phosphorylated following ionizing radiation resulting in Cdc25A instability.[168,169,171-173] Mutation of S76 to alanine stabilizes human Cdc25A [171,172]; however, neither S76A nor S124A overrides checkpoint arrest following ionizing radiation or UV.[168] Mouse cells homozygous for the Cdc25A S124A mutation display Cdc25A stabilization following ionizing radiation;[174] however, there are no changes in proportion of cells in S-phase, or radiation resistant DNA synthesis indicating their S-phase checkpoint is intact.

Human Cdc25A S76, S124 and/or S178 are identified in several publications as "S75, S123 and S177," respectively.[168,169,174-176] In the original cloning of human Cdc25A.[37], there are several substitutions in the N-terminus (Accession: AAA58415.1) and a one residue gap corresponding to residue R12 in all other full length human Cdc25A sequences in the NCBI database (Accession: P30304). Residues have been re-numbered as per "P30304" for the sake of consistency.

Cdc25A S76, S124, S178 and T507 match the consensus site for 14-3-3 binding, RXX$_p$S/T [177] However, only the Chk1 dependent phosphorylation of S178 and T507 results in association with 14-3-3.[170] Substitution of these residues to alanine results in a complete loss of 14-3-3 interaction *in vivo*. Phosphorylation of T507 in particular, and subsequent 14-3-3 binding, interferes with Cdk1-cyclin B association by blocking a cyclin B docking site. A recent study suggests that a ternary complex between Cdc25A, 14-3-3γ and Chk1 is formed following ionizing radiation.[178] The Chk1/14-3-3γ interaction requires auto-phosphorylation of Chk1 S296. Substituting Chk1 S296 for alanine precludes 14-3-3 binding and Cdc25A S76 phosphorylation and deactivates the DNA damage checkpoint.[178] 14-3-3 proteins preferably exist as thermostable homo- and heterodimers. Each isoform in a heterodimer binding a different protein provides a common mechanism for bringing enzymes and their substrate proteins into close proximity.[179,180]

Defects in the Intra-S checkpoint allow replication of damaged DNA.[173,176,181] In mammalian cells, DNA damage results in destabilization of Cdc25A and inhibition of Cdk2.[173] Cdk2 is involved in loading the Cdc45 origin binding factor. Inhibition of Cdc25A stops further origin firing once DNA damage is detected.[182] Cdc25A is also unstable after HU induced replication fork stalling, which unlike in fission yeast, is controlled by activation of Chk1 in mammalian cells.[183]

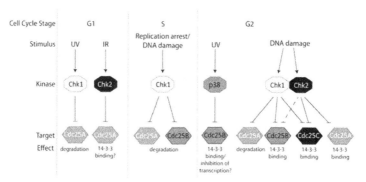

Figure 2. Inhibitory phosphorylation of Cdc25 family members following DNA damage and replication arrest

5.2.1.1. Phosphorylation mediated degradation of Cdc25A by the SCF- βTrCP complex

The mechanism by which Cdc25A is destabilized following DNA structure checkpoint activation is well understood (Figure 3). Human Cdc25A phosphorylation by Chk1 following S-phase DNA damage causes degradation by F-box protein β-TrCP associated with the Skp1-cullin-Fbox (SCF) complex.[171,184] Mutating the destruction box or KEN box (APC interaction motifs) of Cdc25A does not stabilize the protein following exposure to ionizing radiation.[152] This indicates that SCF-mediated degradation of Cdc25A after DNA damage is independent of the cell cycle regulated APC-mediated degradation which occurs at mitotic exit. β-TrCP recognizes a degron motif of DSG(X)₄S where both serine residues are phosphorylated.[185] siRNA knockdown of β-TrCP causes stabilization of Cdc25A, and radiation-resistant DNA synthesis in cells exposed to ionizing radiation.[171,184] Human Cdc25A contains such a motif: DS$_{82}$GFCLDS$_{88}$. The S88A substitution does not stabilize the phosphatase, suggesting that S88 phosphorylation is not explicitly required for β-TrCP binding to Cdc25A.[171] Following ionizing radiation, Chk1 primes Cdc25A for destruction through phosphorylation of S76.[171,186] The NimA-related NEK11 kinase targets Cdc25A S82 and S88, within the DSG degron sequence.[187] Depletion of NEK11 causes a marked decrease in S82 and S88 phosphorylation *in vivo*, and prevents Cdc25A degradation following IR. NEK11, itself thought to be a Chk1 substrate, can directly phosphorylate S82 and S88 *in vitro*.

Following ATM activation, Chk2 phosphorylates and activates the oncogene p53, and inhibits its negative regulator MDM2 after ionizing radiation.[188-191] p53 then induces the Cdk2-cyclin E inhibitor p21^{WAF}.[192] p53 also activates Glycogen Synthase Kinase (GSK-3β), which phosphorylates human Cdc25A S76 following a priming phosphorylation of S80 which can be targeted *in vitro* by Plk3.[193,194] Plk3 is unique among Polo kinase family members in causing Cdc25A stabilization when knocked down with siRNA.[194] In Plk3 null mice Cdc25A is stabilized following DNA damage.[195] Plk3 also phosphorylates and activates p53 following DNA damage and contributes to Chk2 activation.[196,197] In HeLa cells, Casein Kinase 1α (CK1α) sequentially phosphorylates both S79 and S82 following priming phosphorylation of S76 by Chk1 or GSK-3β.[198] The CK1ε isoform negatively regulates Cdc25A by S82 phosphorylation in HEK293 cells.[199] S82 phosphorylation *in vitro* and *in vivo* takes place in response to DNA damage, dependent on CK1α first phosphorylating S79.[198] S79A substitution prevents S82 phosphorylation. S76 phosphorylation appears to be the first target in the cascade, since it does not require prior phosphorylation of S79, T80, S82 or S88. Only alanine substitutions of S76, S79 and S82 impair β-TrCP interaction with Cdc25A and stabilize the phosphatase suggesting that T80 and S88 are not critical for Cdc25A degradation. Additional sites (S107, S156, S192, S279 S293) are phosphorylated by Chk1 or Cds1 *in vitro* but mutating any of these sites does not eliminate β-TrCP binding.[200] Although an interaction between phosphorylated *Xenopus* Cdc25A and β-TrCP has not been demonstrated, Chk1 is required for Cdc25A degradation at the mid-blastula transition through phosphorylation of S73.[201] Interestingly, *Xenopus* Cdc25A lacks the first serine in the DSG(X)₄S degron motif found in human Cdc25A, instead possessing DDG. *Xenopus* Cdc25A S73 lies just upstream of the mutated degron and is analogous to S76 in human Cdc25A. The DAG motif of *Xenopus* Cdc25A lies within a larger PEST motif which regulates stability of the phosphatase through β-TrCP binding, independent of Chk1.[202] In mouse embryonic stem cells, which lack a G1/S DNA damage response, Cdk2 kinase activity is not affected by Cdc25A degradation following exposure to ionizing radiation.[203] Cdc25A degradation is independent of Chk1 and Chk2 but instead dependent on GSK-3β. Like Cdc25A, mammalian Cdc25B is unstable following HU induced arrest.[117] Cdc25B binds β-TrCP strongly and contains the residues DAG rather than the DSG degron motif.[202,204] In contrast, Cdc25B accumulates following G2 DNA damage checkpoint arrest induced by a variety of agents.[205] This is reminiscent of the "stockpiling" phenomena noted earlier in *S. pombe*.[70] β-TrCP interaction with Cdc25B may also be required for mitotic exit.[204] A Cdc25B-DDA degron mutant which cannot bind β-TrCP accelerates mitotic entry slightly, but has a significant delay completing mitosis and progressing to G1. This mitotic delay is due to an extended metaphase in which Cdc25B-DDA shows a high proportion of lagging, misaligned and bridged chromosomes as well as mis-oriented spindles.[204]

Figure 3. Multi-step phosphorylation cascades involved in targeting human Cdc25A for degradation by the β-TrCP SCF complex following DNA damage in S-phase and G2.

5.2.1.2. Cdc25A inhibition by 14-3-3 binding

Cdc25A S76, S124, S178 and T507 match the consensus site for 14-3-3 binding RXX$_p$S/T [177] However, only the Chk1 dependent phosphorylation of S178 and T507 results in association with 14-3-3.[170] Substitution of these residues to alanine results in a complete loss of 14-3-3 interaction *in vivo*. Phosphorylation of T507 in particular, and subsequent 14-3-3 binding, interferes with Cdk1-cyclin B association by blocking a cyclin B docking site. A recent study suggests that a ternary complex between Cdc25A, 14-3-3γ and Chk1 is formed following ionizing radiation.[178] The Chk1/14-3-3γ interaction requires auto-phosphorylation of Chk1 S296. Substituting Chk1 S296 to alanine precludes 14-3-3 binding and Cdc25A S76 phosphorylation and deactivates the DNA damage checkpoint.[178] 14-3-3 proteins preferably exist as thermostable homo- and heterodimers. Each isoform in a heterodimer binding a different protein provides a common mechanism for bringing enzymes and their substrate proteins into close proximity.[179,180]

Following exposure to ionizing radiation during G1 phosphorylation of Cdc25A on S124 by Chk2 prevents entry to S-phase .[173] Chk2 cannot efficiently phosphorylate Cdc25A on S76 and so cannot induce Cdc25A degradation by the β-TrCP route.[206] The mechanism by which S124 phosphorylation induces Cdc25A degradation is not clear because it is not required for degradation via β-TrCP.[173] S124 conforms to a 14-3-3 phospho-serine binding site, but doesn't bind 14-3-3.[170]In contrast to the effects of Cdc25A phosphorylation sites discussed thus far, modification of *Xenopus* Cdc25A T504 by Chk1 negatively regulates interaction with Cdk1-cyclin A, Cdk1-cyclin B and Cdk2-cyclin E but in a 14-3-3 independent manner.[207]

5.2.2. G2/M DNA damage checkpoint

The response to damage in G2 is dependent on the nature of the damage signal. Exposure to UV activates p38 MAP kinase and checkpoint arrest is independent of ATM and ATR since the arrest is not caffeine sensitive.[208] The primary target following UV irradiation is Cdc25B and although p38 can phosphorylate Cdc25C *in vitro*, UV exposure does not affect the Cdc25C/14-3-3 interaction.[208] Ionizing radiation activates ATM and ATR and results in Cdc25C phosphorylation by Chk1 and Chk2.[209]

5.2.2.1. Cdc25B inhibition following UV induced DNA damage

A number of conflicting reports have appeared relating to Cdc25B regulation following UV exposure. Some groups have reported that UV has no effect on Cdc25B protein levels [208,210], but others have shown that UV causes either MAP kinase mediated Cdc25B degradation or Cdc25B accumulation.[205,208,211] Cell line specific effects have no doubt contributed to these inconsistencies as human molecular biology relies heavily on transformed cell lines and mis-regulation of Cdc25B is a common phenomenon in tumors.[212] Cdc25B isolated from UV irradiated A2058 melanoma cells still retains a substantial portion of its Y15 phosphatase activity and is localized to the nucleus.[213] In HeLa cells Cdc25B is localized almost exclusively to the cytoplasm as detected by cell fractionation and immunofluorescence.[120] Variation in the apparatus used for UV irradiation could also have contributed to contradictory accounts of Cdc25B regulation. A recent re-examination of the effect of UV on Cdc25B showed that after 10 J/m² exposure, Cdc25B levels did not decrease, although following 60 J/m² Cdc25B was clearly downregulated.[214] Exposure of U2OS osteosarcoma cells to 10 J/m² UV leads to Cdc25B nuclear export. Based on chemical inhibitor experiments Cdc25B downregulation is not mediated by ATM/ATR, p38 MAPK or JNK, but rather following 60 J/m², by inhibiting Cdc25 translation. The eukaryotic initiation factor regulating Cdc25B expression, eIF2α, is phosphorylated and inhibited following UV exposure.[215] UV mediated DNA damage during G2 involves human Cdc25B S323 phosphorylation through the p38 kinase during interphase.[208] ATM/ATR inhibitor caffeine and the Chk1 inhibitor UNC-01 have no effect on UV mediated checkpoint arrest. Isoforms 14-3-3β and 14-3-3ε bind preferentially to Cdc25B phosphorylated S323, allowing its nuclear export.[216] Nuclear export of Cdc25B is abolished in cells expressing the Cdc25B S323A substitution, regardless of which 14-3-3 isoform is co-expressed.[216] Two amino-truncated Cdc25B isoforms localize to the nucleus *in vivo* and regulate recovery from G2/M checkpoint arrest but neither is required for mitotic entry.[44] Although one of these isoforms contains the DDG degron described above, both are more stable than the full length Cdc25B.

5.2.2.2. Inhibition of Cdc25C following DNA damage

Cdk1 activation is prevented by UV induced checkpoint activity coincident with the appearance of a phosphorylated form of Cdc25C.[213] In contrast with fission yeast Cdc25, which gradually accumulates in the nucleus during G2, human Cdc25C is primarily localized to the cytoplasm during interphase and only enters the nucleus at mitotic entry.[217] Thus, nuclear export of Cdc25C is not a requirement for G2 DNA damage response, since the phosphatase is already cytoplasmic at this time. However, exposure to ionizing radiation decreases the enzymatic activity of human Cdc25C.[218] Several research groups showed relatively early in the Cdc25 phosphorylation story that residue S216 is phosphorylated by Chk1 [219-221] and Chk2 *in vitro*.[209,222] Phosphorylation of S216 results in 14-3-3 binding and nuclear export.[219] Residues surrounding S216, RSPS$_{216}$MP, correspond to the canonical 14-3-3 consensus binding site RSX$_p$SXP.[177] Cdc25C S216A mutants are unable to bind 14-3-3.[219] This is likely due to close proximity of S216 to the NLS, leading to the obstruction of the import signal and trapping of the phosphatase in the cytoplasm.[221,223] A higher proportion of cells expressing Cdc25 S216A have nuclear abnormalities indicative of progression to mitosis prior to completion of DNA synthesis; this effect is exacerbated by addition of HU.[217,219] Such cells also have reduced ability to delay entry to mitosis following ionizing radiation exposure.[224]

Although Cdc25C S216A is a relatively poor substrate for Chk1 compared to the wildtype protein, Chk1 can still execute some degree of phosphorylation on the mutant phosphatase.[220] This observation suggests the possibility that additional Cdc25C phosphorylation negatively regulates its enzymatic activity. Recent bioinformatics approaches to generate profiles from peptide binding arrays based on three diverse 14-3-3 binding sites have generated an improved 14-3-3 binding motif consensus.[225] This helped to identify two additional phosphorylated Cdc25C residues, Ser247 and Ser263, which interact with 14-3-3. Mutation of either residue to alanine reduces 14-3-3 binding, but neither of these mutant peptides was affected by Cdc25 S216 phosphorylation when expressed in cells. S263 was previously identified in an isolated report which showed phosphorylation of this residue induces Cdc25B nuclear export.[226] Cdc25C purified from cells treated with the topoisomerase II inhibitor etoposide is phosphorylated on S263, but S263A substitution results in enhanced nuclear localization. The kinase targeting this residue has yet to be determined experimentally. However, the residues surrounding S263 (KKTV$_p$SLCD) conform to a Chk1 consensus site as Chk1 can tolerate a lysine (K) at the -3 position relative to the phosphorylated serine or threonine *in vitro*.[227]

Regulation of *Xenopus* Cdc25C localization is similar to its human counterpart. In cultured *Xenopus* tissue Cdc25C is primarily in the cytoplasm while the Cdc25C S287A mutant, corresponding to S216 in human Cdc25C, is almost exclusively nuclear.[228] Cdc25C is phosphorylated on multiple sites by Chk1 but only S287 phosphorylation is required for 14-3-3 binding.[229-231] Although *Xenopus* Cdc25C can be made exclusively cytoplasmic by co-expression with 14-3-3ε, Cdc25C S287A is not affected. Nuclear export depends on the intrinsic 14-3-3ε nuclear export signal and re-import is prevented by blocking Cdc25C association with importin-α.[228] In egg extracts depleted of endogenous Cdc25C, expressing Cdc25C S287A accelerates mitotic entry relative to overexpression of the wildtype protein.[229] However, Cdc25C S287A has a less than two fold increase of *in vitro* Cdk1 Y15 dephosphorylation activity over wildtype Cdc25C when pre-incubated with 14-3-3ε. Thus, it seems likely that *Xenopus* Cdc25C phosphorylation and 14-3-3 binding regulates the phosphatase at the level of cellular localization, rather than inhibiting its activity *per se*.

Cdc25A is considered to regulate the G1/S transition in the "Traditional Model" of the human cell cycle but it also has a significant role in mitotic entry. As such, Cdc25A is an important target of the DNA damage checkpoint. In fact, mice lacking Cdc25B and Cdc25C do not have a G2/M checkpoint defect.[106,107] Phosphorylation by Chk1 causes degradation of Cdc25A following DNA damage during G2 ionizing radiation and exposure to the DNA intercalating agent adriamycin.[172,181]

5.3. Chk1 regulation of Cdc25 orthologues in unperturbed cell cycles

Cdc25A, B and C are all phosphorylated by Chk1 in the absence of externally induced DNA damage. As Cdc25 phosphatases are such potent positive regulators of cell cycle transitions it is perhaps not surprising that the cell maintains their activity at a low level until their precise point of activation. Chk1 regulates human Cdc25A stability during unperturbed cell cycles. Phosphorylation of S82 and S88 can be detected using phospho-specific antibodies in unperturbed cells.[184] Depletion of the S82/S88 kinase NEK1 and S76 kinase CK1ε by siRNA, results in Cdc25A stabilization in the absence of DNA damage.[187,199] Cdc25A S124 phosphorylation by Chk1 also occurs in the absence of damage and destabilizes the phosphatase.[168,169,200] Inhibiting ATM/ATR, or a variety of upstream checkpoint components, also stabilizes in the absence of externally induced DNA damage which may indicate there is some basal level of spontaneous damage checkpoint signaling.[232] Cdc25A is also phosphorylated by Casein kinase 2β (CK2β) in a damage independent manner.[233] CK2 phosphorylates human Cdc25C T236 adjacent to the NLS *in vitro*.[234] A T236D mutation reduces β-importin binding, thus excluding Cdc25 from the nucleus.[234] CK2 phosphorylates Cdc25B on residues S186 and S187, just downstream from the KEN box, modestly increases its phosphatase activity, and potentially blocking APC mediated degradation.[235]

Human Cdc25B is phosphorylated *in vitro* by Chk1 at S230 and S563 in the absence of DNA damage.[236] S230 phosphospecific antibodies show that Cdc25B modified on this residue is centrosome associated from S-phase until mitosis.[236] A population of Chk1 is localized to the centrosome during G2 and can prevent promiscuous Cdk1-cyclin B activation by Cdc25B.[237] S323 was previously identified as the major 14-3-3 binding site.[238] S151 and S230 account for the remainder of the interaction, but both need to be dephosphorylated before interaction with 14-3-3 is lost.[239] Human Cdc25B S563 resides in the extreme C-terminus of Cdc25B and is analogous to Cdc25A S504 and *S. pombe* Cdc25 T569. Cdc25B lacking the S323 phosphorylation site is almost exclusively nuclear. It is interesting that this residue is targeted by Chk1, but does not appear to bind 14-3-3 when phosphorylated. If the function of this phosphorylation is conserved between Cdc25A S504 and Cdc25B S563, phosphorylation may affect interaction with Cdk1-cyclin B.[240]

Chk1 phosphorylation of human Cdc25C and nuclear export by 14-3-3 binding keeps the phosphatase cytoplasmic during unperturbed cell cycles.[217] Nuclear localization of Cdc25C(S216A) is enhanced, suggesting that part of the function of 14-3-3 binding is to obscure the NLS located adjacent to this residue.[224] Overexpression of Cdc25C(S216A) induces a higher degree of premature mitotic entry.[217] *Xenopus* Cdc25C is likewise phosphorylated on S287 by Chk1 during interphase.[228,241] Phospho-S287 is bound by 14-3-3ε and 14-3-3ζ obscuring the NLS and preventing nuclear import.[228] *Xenopus* Cdc25C is phosphorylated on T533 by Chk1, but not Chk2.[207] Injecting Cdc25C T533A mRNA into *Xenopus* oocytes results in more rapid dephosphorylation of Cdk1 Y15.[207] Again, this suggests that regulation of CDK-cyclin interaction with Cdc25 orthologues by C-terminal phosphorylation is a common mechanism for inhibiting cell cycle progression.

Although Cdc25A and Cdc25B are dispensable for embryonic development, Chk1 and ATR kinases are essential.[242,243] The Cdc25B/14-3-3 interaction is important for maintaining G2 arrest, and inhibiting germinal vesicle breakdown in *Xenopus* oocytes prior to progesterone exposure.[244]

In *S. pombe*, Chk1 does not appear to negatively regulate Cdc25 in the absence of DNA damage. Loss of a negative regulator of Cdc25 is expected to cause the cell to divide at a reduced length. However, deletion of Chk1 does not cause a cell cycle phenotype.[245] In addition, expressing Cdc25 where all twelve putative Cds1/Chk1 phosphorylation sites are mutated to alanine does not cause acceleration of the cell cycle.[60]

5.4. Cdk1 phosphorylation of Cdc25 precludes checkpoint mediated inhibition during mitosis

If Cdc25C is phosphorylated and inactivated during interphase via 14-3-3 binding, how is it then activated at mitotic entry? Cdk1-cyclin B phosphorylation of Cdc25C causes 14-3-3 dissociation and allows removal of interphase phosphorylations. Re-phosphorylation of these residues is simultaneously blocked. Cdk1-cyclin B thus potentiates Cdc25C for its pro-mitotic function and ensures that it remains active. The region surrounding S216 in human Cdc25C and S287 in *Xenopus* Cdc25C is a well conserved stretch in the N-terminal region of the two proteins; LYRSPS$_{216}$MPE is identical between Human and *Xenopus* and contains S216 and S287, respectively.(Figure 4) In both organisms, the two serine residues upstream of the major phosphorylated 14-3-3 binding residue, S214 in human and S285 in *Xenopus*, is targeted by Cdk1-cyclin B.[42,246-248] In human cells phosphorylation of S214 precludes phosphorylation of S216 by Chk1 and 14-3-3 binding.[42,246] Substituting S214D prevents phosphorylation of S216.[246] S216 is not phosphorylated during M-phase *in vivo*, and ionizing radiation in M-phase cells cannot induce its phosphorylation.[246] 14-3-3 is unable to bind Cdc25C immunoprecipitated from cells arrested in mitosis with nocodazole.[219] In *Xenopus*, Cdc25C S285 phosphorylation prevents the phosphorylation of S287.[247,249] S285 can be phosphorylated by both Cdk1-cyclin B and Cdk1-cyclin A *in vitro*.[146] Until the mid-blastula transition, *Xenopus* Cdc25C is phosphorylated on S285 which precludes Chk1 mediated phosphorylation of S287 and subsequent 14-3-3 binding.[247] Human Cdc25C S214 is also phosphorylated in maturing human oocytes.[250] Removal of phospho-S287 bound 14-3-3 and phosphorylation of *Xenopus* Cdc25C S285 requires prior phosphorylation of Cdc25C T138.[249] T138 is a substrate of Cdk1-cyclin B.[146] However, selective depletion of Cdk2 from *Xenopus* egg extracts using the N-terminus of the Cdk2 inhibitor p21, completely prevents removal of 14-3-3 from Cdc25C.[251] In *Xenopus* egg extracts Cdk2 is required for mitotic entry.[81] Other CDK phosphorylation sites on *Xenopus* Cdc25C such as T48 and T67 negatively regulate its activity in an S287 independent manner.[249] Phosphorylation of T138 is not sufficient to cause 14-3-3 dissociation.[248] How is 14-3-3 physically removed from phospho-S287? The pelleted fraction from ultracentrifuged interphase egg extracts contains a 14-3-3 dissociating activity, which was determined to be the intermediate filament component Keratin 8/18[248]. Keratin is phosphorylated during mitosis, binds 14-3-3, and has a role in mitotic progression in hepatocytes.[252] Keratin may act as a "14-3-3 sink" during mitosis, stripping 14-3-3 from S287.[248] *Xenopus* Cdc25C T138 corresponds with T130 in human Cdc25C. A phospho-T130 specific antibody shows that Cdc25C phosphorylated on this site localizes to the centrosome.[145] A localization for the de-inhibition of Cdc25C fits well with the putative centrosomal localization of Cdc25C activation.

PP1 phosphatase removes S287 phosphorylation once 14-3-3 has dissociated.[251] Binding of PP1 to *Xenopus* Cdc25C requires a docking motif "VXF", amino acids 105-107, the loss of which prevents S287 dephosphorylation.[251] Phosphorylation of *Xenopus* Cdc25C S285 by Cdk1-cyclin B enhances recruitment of phosphatase PP1 to Cdc25C, inducing the dephosphorylation of S287.[249] PP2A/B56δ dephosphorylates *Xenopus* Cdc25C T138 during interphase, mitotic exit and following replication arrest.[248] This maintains the phosphatase in a state where it can be inhibited by S287 phosphorylation. Inhibition of PP2A by okadaic acid prematurely induces mitotic entry in *Xenopus* egg extracts.[253] T138 is also dephosphorylated during replication arrest where B56δ is itself phosphorylated by Chk1, enhancing complex formation with PP2A.[248] PP2A mediated dephosphorylation of Cdc25 is assisted by the action of the Pin1 prolyl isomerase. Pin1 isomerizes the peptide bond between phospho-serine/threonine and proline placing their R-groups in a *trans* orientation.[254] This isomerization makes the phosphorylated residue a better substrate for the PP2A phosphatase.[255] *In vitro*, Pin1 decreases the catalytic activity of Cdc25C that has previously been phosphorylated by Cdk1.[256] Pin1 is also involved in maintenance of replication checkpoint arrest in *Xenopus* encouraging the reversal of Cdk1 mediated Cdc25C phosphorylation.[257] In Humans, Cdc25C is deactivated in a similar manner. T130 on Cdc25C is dephosphorylated by PP2A.[258] Mitotic exit is delayed when PP2A is knocked down, suggesting that dephosphorylation of Cdc25C T130 in human cells is also important for the transition from M to G1 of the next cycle. A similar situation may exist where human Cdc25B1

S321 phosphorylation blocks phosphorylation of the major 14-3-3 binding residue S323 by p38.[246] Expression of the phosphorylation-mimicking Cdc25B1 S321D prevents p38 mediated phosphorylation of S323 and abolishes binding by 14-3-3β and 14-3-3ε.[246,259] 14-3-3σ binding is unaffected by S321 phosphorylation.[259] 14-3-3σ preferentially interacts with Cdc25B3 S230 [216] and is induced by p53 following activation by ATM/Chk2.[260]

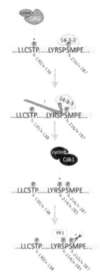

Figure 4. Cdk-cyclin mediated removal of inhibitory S216/S287 phosphorylation and 14-3-3 binding in human and *Xenopus* Cdc25C, respectively.

There are no Cdc2 phosphorylation motifs (S/TP) directly upstream of any of the twelve Cds1 *in vitro* phosphorylation sites in *S. pombe*. Large scale phosphoproteome analysis has detected S99, S178 and S359 phosphorylation of fission yeast Cdc25 in M-phase arrested cells.[261] Thus it appears that the vertebrate mechanism for reversing and preventing Chk1 phosphorylation of Cdc25C evolved relatively recently. However, parallels exist between some aspects of Cdc25 dephosphorylation between fission yeast and vertebrates. Treatment with okadaic acid or deletion of PP2A homologue Ppa2 causes premature mitotic entry in fission yeast.[262,263] Loss of *ppa2* suppresses the *cdc25-22ts* mutation, a genetic interaction indicative of a negative regulator.[263] Loss of PP1 homologue Dis2 results in cell elongation, suggesting its role as a positive regulator of Cdc25 is conserved in *S. pombe*.[264] Temperature sensitive *dis2* mutants have a defect in exit from mitosis, similar to the effect of siRNA inhibition of human PP2A.[258,264]

6. Cdc25 phosphorylation by MAP kinase cascades

In addition to regulation by DNA damage and DNA replication checkpoints, Cdc25 is the target of several MAPK cascades responding to stress and mitogenic signals. A detailed description of the variety of MAPK pathways is outside of the scope of this manuscript, but excellent reviews are available.[265]. In general, MAP kinase cascades involve the sequential activation of three kinases; a MAP kinase kinase kinase (MAPKKK) phosphorylates a MAP kinase kinase (MAPKK), which phosphorylates a MAP kinase (MAPK). There are three such cascades which are salient to our discussion of Cdc25 regulation.

6.1. Raf/MEK/ERK

The ERK1/ERK2 MAPKs are activated by Raf MAPKKKs working on MEK1/MEK2 MAPKKs. This cascade is primarily activated by extracellular signaling through receptor tyrosine kinases in response to mitogenic signals.[266] During Cyto-Static Factor arrest in mature *Xenopus* oocytes metaphase II arrest is enforced by the combined activity of the MEK/ERK pathway upregulating ribosomal subunit S6 kinase/CamKII homologue p90^{rsk}, and APC inhibition by the Emi2 kinase.[267] Calcium influx as a result of fertilization activates p90^{rsk} (a Ca^{2+}/calmodulin dependent kinase II homologue) which phosphorylates Emi2.[268] This allows APC activation, cyclin B degradation, and progression through meiosis II. Cdc25C is phosphorylated on S287 by p90^{Rsk} *in vitro* resulting in its inhibition via 14-3-3 binding.[269,270] Conversely, *Xenopus* Cdc25C S317, T318 and S319 are phosphorylated by p90^{rsk} orthologue Rsk2 which serves to activate the phosphatase in mature oocyte extracts.[271] The *Xenopus* ERK homologue, p42, phosphorylates Cdc25C T48, T138 and S205 *in vitro*.[272] T138 and S205 are also Cdk1-cyclinB targets [146], but T48 is uniquely phosphorylated by p42. In HeLa cells after activation of MEK/ERK signaling by addition of 12-O-tetradecanoylphorbol-13- acetate (TPA), G2/M transition is inhibited by degradation of Cdc25B.[273] This is mediated through MEK dependent phosphorylation of Cdc25B S249 by CamKII, an activator of human Cdc25C.[274] Inhibition of CamKII results in a G2 arrest. Overexpression of CamKII can also arrest cells in G2, but with a high level of Cdk1/Cyclin B activity.[275] Another member of the CamKII family, C-TAK1, phosphorylates human Cdc25C on S216 resulting in inhibition via 14-3-3 binding and nuclear export.[276] *Xenopus* Cdc25C can be phosphorylated by CamKII *in vitro*.[277] Lastly, p14^{arf} upregulation in response to anti-proliferative signals results in human Cdc25C downregulation through MEK/ERK MAPK signaling.[278] Human Cdc25C becomes phosphorylated on S216 and is subsequently ubiquitinated and degraded.[278]

6.2. p38 and JNK MAPKs

The p38 and JNK kinases activate in response to extracellular stimuli such as heatshock, oxidative stress, ionizing radiation, UV and growth factor deprivation. Both are activated by MAPKKs of the MKK family which are themselves activated by a large variety of MAPKKKs. We have already discussed the function of p38 in response to UV irradiation. p38 also phosphorylates human Cdc25A on S124 and S76 in response to osmotic stress, destabilizing the protein [168] and a 42 C heatshock causes p38 and Chk2 to phosphorylate S76 and S178 of human Cdc25A, respectively.[175] MAPKAP kinase 2 (MK2) functions downstream of p38 and regulates G1 and S-phase cell cycle progression in response to UV.[210] Downstream of p38, it phosphorylates RxxS/T motifs and activation of MK2 correlates with increased binding of 14-3-3 to Cdc25B. p38 and MK2 kinase form a tight complex and are imported into the nucleus together, so previous work showing that p38 directly phosphorylates S216 on Cdc25C and S323 on Cdc25B may in fact have been inadvertently monitoring MK2 activity.[279] Cdc25A may also be an MK2 substrate as MK2 knockdowns ablate the G1 and S-phase checkpoints. JNK activity targets two serines within the region DAGLCMDS$_{101}$PS$_{103}$P of the DSG degron on human Cdc25B.[280] Simultaneous S101A and S103A substitution prevents β-TrCP binding and Cdc25B ubiquitination. JNK also phosphorylates Cdc25C on S168 inhibiting its phosphatase activity.[281,282] This residue is transiently phosphorylated *in vivo* on nuclear Cdc25C prior to and after mitotic entry.[281] S168 is phosphorylated following UV irradiation and osmotic shock.[281,282]

6.3. The *S. pombe* stress activated MAP kinase pathway

In fission yeast, Cdc25 is phosphorylated by the CamKII homologue Srk1 in response to extracellular stress.[283] Srk1 is activated downstream of the Spc1 MAPK, Wis1 MAPKK, and Win1 or Wak1MAPKKKs.[284] This phosphorylation occurs on residues also targeted by Cds1 as Cdc25(9A) is not sensitive to Srk1 mediated inhibition.[283] Srk1 phosphorylation of Cdc25 results in its nuclear export, similar to the response to DNA damage and replication arrest.

7. Other kinases which phosphorylate Cdc25s

PKA prevents oocyte maturation by inhibition of Polo kinase mediated Cdc25 activation, and deactivating the Mos/MEK/ERK MAP kinase cascade which inhibits Myt1.[285] Progesterone exposure in *Xenopus* oocytes down regulates adenylate cyclase, lowering cyclicAMP levels and consequently deactivating PKA. Prior to maturation, *Xenopus* Cdc25C is also inhibited by PKA phosphorylation of S287 and T318 *in vitro*.[286] Murine PKA phosphorylates Cdc25B S321, negatively regulating the phosphatase; S321A mutants cause enhanced germinal vesicle breakdown when injected into mouse oocytes.[287]

Pim1 is a serine/threonine kinase induced by the SAT3 and SAT5 transcription factors following cytokine exposure thus linking pro-proliferative signals to the cell cycle control machinery.[288] Pim1 phosphorylates and activates Cdc25A and represses the Cdk4/6 inhibitor p21^{CIP} to encourage the G1/S transition.[289,290] Pim1 is also able to phosphorylate and inhibit the CamKII homologue c-TAK and accelerate Cdc25C mediated mitotic entry.[291]PAR-1/MARK (partitioning-defective 1/Microtubule affinity-Regulating Kinase) protein homologue pEG3 phosphorylates human Cdc25A S263, Cdc25B S169 and Cdc25C S216.[292,293] Overexpressing pEg3 results in G2 arrest which can be reversed by co-expressing Cdc25B.[292] Anti-phospho PAR1 S169 antibodies stain spindle pole and centrosome in immunofluorescence experiments.[293] In *C. elegans* the first cell division is unequal which produces a larger anterior cell and a smaller posterior one. The next cell division is asynchronous with the anterior cell dividing prior to the posterior one.[294] Rapid anterior cell cycle timing is due to enrichment of Polo kinase and Cdc25.1 in the anterior cell. This is dependent on a network of polarity proteins, including Par1.[295]

8. Cdc25 and disease

Cdc25 orthologues are the subject of much attention as they are commonly upregulated in human tumors.[296] This is perhaps not surprising considering the role of Cdc25 inhibition in maintaining genomic stability and the regulation of these phosphatases by Rb, p53 and a number of other oncogenes. Cdc25A and Cdc25B themselves are oncogenes in humans.[212] Cdc25B is overexpressed in 32% of breast cancer tissue samples, and high Cdc25B levels correlate with high incidences of recurrence and decreased 10 year survival.[212] Overexpression of Cdc25A is similarly linked to poor clinical outcome.[296,297] Cells bearing oncogenic mutations of myc have elevated Cdc25A and Cdc25B levels.[298] Anti-Cdc25B autoantibody has been shown to be a predictor of poor prognosis in esophageal cancer patients.[299] Overexpression of Cdc25B has recently been shown to cause a variety of S-phase effects including increased Cdc45 recruitment to chromatin, impairment of replication fork progression DNA damage and chromosome instability.[300]

An interesting link between Cdc25 and disease comes from the finding that the HIV-1 protein vpr causes G2/M arrest.[301] When expressed in *S. pombe*, vpr activates Srk1 kinase, resulting in Cdc25 phosphorylation and 14-3-3 mediated nuclear export.[302,303] The vpr protein also acts through upregulation of PP2A phosphatase acting on both Wee1 kinase and Cdc25, reversing activating Cdc2 phosphorylation.[304]

9. Conclusion

It has been more than thirty five years since Cdc25 was first isolated as an elongated temperature-sensitive fission yeast mutant and twenty one years since its biochemical function was determined. The field of cell cycle research and the study of Cdc25 in particular are extremely active with numerous new manuscripts appearing each year. This research has revealed that Cdc25 is one of the most intricately regulated proteins in the cell. Cdc25 accepts input from numerous pathways and checkpoints monitoring whether conditions inside and outside the cell are permissive for cell cycle progression. When conditions warrant caution, Cdc25 is inhibited by phosphorylation leading to alterations in its catalytic activity, cellular localization, substrate recognition and stability. When the green light is given Cdc25 participates in an intricate series of interconnected positive feedback loops with the beating heart of cell cycle regulation, the Cyclin-CDK complex. When the cell loses control of Cdc25 regulation, the results are deadly.

Author details

C. Frazer
Department of Biomedical and Molecular Science, Queen's University, Kingston, Ontario, K7L 3N6, Canada

P.G. Young*
Department of Biology, Queen's University, Kingston, Ontario, Canada

---

* Corresponding Author

Acknowledgement

Thank you to Silja Freitag for her critical reading of this manuscript. This work was supported by a grant from the Natural Sciences and Engineering Research Council of Canada to PGY.

10. References

[1]   Hartwell LH, Culotti J, Reid B. (1970) Genetic control of the cell-division cycle in yeast. I. Detection of mutants. Proc Natl Acad Sci U S A 66(2):352-359.

[2]   Hartwell LH, Mortimer RK, Culotti J, Culotti M. (1973) Genetic control of the cell division cycle in yeast: V. Genetic analysis of *cdc* mutants. Genetics 74(2):267-286.

[3]   Nurse P, Thuriaux P, Nasmyth K. (1976) Genetic control of the cell division cycle in the fission yeast *Schizosaccharomyces pombe* . Mol Gen Genet 146(2):167-178.

[4]   Nurse P. (1975) Genetic control of cell size at cell division in yeast. Nature 256(5518):547-551.

[5]   Fantes P. (1979) Epistatic gene interactions in the control of division in fission yeast. Nature 279(5712):428-430.

[6]   Fantes PA. (1981) Isolation of cell size mutants of a fission yeast by a new selective method: characterization of mutants and implications for division control mechanisms. J Bacteriol 146(2):746-754.

[7]   Thuriaux P, Nurse P, Carter B. (1978) Mutants altered in the control co-ordinating cell division with cell growth in the fission yeast *Schizosaccharomyces pombe*. Mol Gen Genet 161(2):215-220.

[8]   Kishimoto T, Kanatani H. (1976) Cytoplasmic factor responsible for germinal vesicle breakdown and meiotic maturation in starfish oocyte. Nature 260(5549):321-322.

[9]   Masui Y, Markert CL. (1971) Cytoplasmic control of nuclear behavior during meiotic maturation of frog oocytes. J Exp Zool 177(2):129-145.

[10] Maller JL. (1990) Xenopus oocytes and the biochemistry of cell division. Biochemistry 29(13):3157-3166.

[11] Smith LD, Ecker RE. (1971) The interaction of steroids with Rana pipiens Oocytes in the induction of maturation. Dev Biol 25(2):232-247.

[12] Murray AW, Solomon MJ, Kirschner MW. (1989) The role of cyclin synthesis and degradation in the control of maturation promoting factor activity. Nature 339(6222):280-286.

[13] Newport J, Kirschner M. (1982) A major developmental transition in early Xenopus embryos: II. Control of the onset of transcription. Cell 30(3):687-696.

[14] Newport J, Kirschner M. (1982) A major developmental transition in early Xenopus embryos: I. characterization and timing of cellular changes at the midblastula stage. Cell 30(3):675-686.

[15] Dunphy WG, Brizuela L, Beach D, Newport J. (1988) The *Xenopus* cdc2 protein is a component of MPF, a cytoplasmic regulator of mitosis. Cell 54(3):423-431.

[16] Evans T, Rosenthal ET, Youngblom J, Distel D, Hunt T. (1983) Cyclin: a protein specified by maternal mRNA in sea urchin eggs that is destroyed at each cleavage division. Cell 33(2):389-396.

[17] Gautier J, Minshull J, Lohka M, Glotzer M, Hunt T, Maller JL. (1990) Cyclin is a component of maturation-promoting factor from Xenopus. Cell 60(3):487-494.

[18] Patra D, Dunphy WG. (1996) Xe-p9, a Xenopus Suc1/Cks homolog, has multiple essential roles in cell cycle control. Genes Dev 10(12):1503-1515.

[19] Gautier J, Norbury C, Lohka M, Nurse P, Maller J. (1988) Purified maturation-promoting factor contains the product of a Xenopus homolog of the fission yeast cell cycle control gene cdc2+. Cell 54(3):433-439.

[20] Nurse P, Bissett Y. (1981) Gene required in G1 for commitment to cell cycle and in G2 for control of mitosis in fission yeast. Nature 292(5823):558-560.

[21] Beach D, Durkacz B, Nurse P. (1982) Functionally homologous cell cycle control genes in budding and fission yeast. Nature 300(5894):706-709.

[22] Hagan I, Hayles J, Nurse P. (1988) Cloning and sequencing of the cyclin-related cdc13+ gene and a cytological study of its role in fission yeast mitosis. J Cell Sci 91 (Pt 4)(Pt 4):587-595.

[23] Booher R, Beach D. (1988) Involvement of cdc13+ in mitotic control in *Schizosaccharomyces pombe*: possible interaction of the gene product with microtubules. EMBO J 7(8):2321-2327.

[24] Gautier J, Solomon MJ, Booher RN, Bazan JF, Kirschner MW. (1991) Cdc25 is a Specific Tyrosine Phosphatase that Directly Activates P34cdc2. Cell 67(1):197-211.

[25] Millar JB, McGowan CH, Lenaers G, Jones R, Russell P. (1991) p80[cdc25] mitotic inducer is the tyrosine phosphatase that activates p34[cdc2] kinase in fission yeast. EMBO J 10(13):4301-4309.

[26] Doree M, Peaucellier G, Picard A. (1983) Activity of the maturation-promoting factor and the extent of protein phosphorylation oscillate simultaneously during meiotic maturation of starfish oocytes. Dev Biol 99(2):489-501.

[27] Maller J, Wu M, Gerhart JC. (1977) Changes in protein phosphorylation accompanying maturation of Xenopus laevis oocytes. Dev Biol 58(2):295-312.

[28] Mazzei G, Guerrier P. (1982) Changes in the pattern of protein phosphorylation during meiosis reinitiation in starfish oocytes. Dev Biol 91(2):246-256.

[29] Wiblet M. (1974) Protein kinase activities during maturation in Xenopus laevis oocytes. Biochem Biophys Res Commun 60(3):991-998.

[30] Margalit A, Vlcek S, Gruenbaum Y, Foisner R. (2005) Breaking and making of the nuclear envelope. J Cell Biochem 95(3):454-465.

[31] Abe S, Nagasaka K, Hirayama Y, Kozuka-Hata H, Oyama M, Aoyagi Y, et al. (2011) The initial phase of chromosome condensation requires Cdk1-mediated phosphorylation of the CAP-D3 subunit of condensin II. Genes Dev 25(8):863-874.

[32] Glotzer M. (2009) The 3Ms of central spindle assembly: microtubules, motors and MAPs. Nat Rev Mol Cell Biol 10(1):9-20.

[33] Holt LJ, Tuch BB, Villen J, Johnson AD, Gygi SP, Morgan DO. (2009) Global analysis of Cdk1 substrate phosphorylation sites provides insights into evolution. Science 325(5948):1682-1686.

[34] Enserink JM, Kolodner RD. (2010) An overview of Cdk1-controlled targets and processes. Cell Div 5:11.

[35] Russell P, Moreno S, Reed SI. (1989) Conservation of mitotic controls in fission and budding yeasts. Cell 57(2):295-303.

[36] Nagata A, Igarashi M, Jinno S, Suto K, Okayama H. (1991) An additional homolog of the fission yeast cdc25+ gene occurs in humans and is highly expressed in some cancer cells. New Biol 3(10):959-968.

[37] Galaktionov K, Beach D. (1991) Specific activation of cdc25 tyrosine phosphatases by B-type cyclins: evidence for multiple roles of mitotic cyclins. Cell 67(6):1181-1194.

[38] Sadhu K, Reed SI, Richardson H, Russell P. (1990) Human homolog of fission yeast cdc25 mitotic inducer is predominantly expressed in G2. Proc Natl Acad Sci U S A 87(13):5139-5143.

[39] Wickramasinghe D, Becker S, Ernst MK, Resnick JL, Centanni JM, Tessarollo L, et al. (1995) Two CDC25 homologues are differentially expressed during mouse development. Development 121(7):2047-2056.

[40] Okazaki K, Hayashida K, Iwashita J, Harano M, Furuno N, Sagata N. (1996) Isolation of a cDNA encoding the X enopus homologue of mammalian Cdc25A that can induce meiotic maturation of oocytes. Gene 178(1-2):111-114.

[41] Forrest AR, McCormack AK, DeSouza CP, Sinnamon JM, Tonks ID, Hayward NK, et al. (1999) Multiple splicing variants of cdc25B regulate G2/M progression. Biochem Biophys Res Commun 260(2):510-515.

[42] Bonnet J, Mayonove P, Morris MC. (2008) Differential phosphorylation of Cdc25C phosphatase in mitosis. Biochem Biophys Res Commun 370(3):483-488.

[43] Wegener S, Hampe W, Herrmann D, Schaller HC. (2000) Alternative splicing in the regulatory region of the human phosphatases CDC25A and CDC25C. Eur J Cell Biol 79(11):810-815.

[44] Jullien D, Bugler B, Dozier C, Cazales M, Ducommun B. (2011) Identification of N-terminally truncated stable nuclear isoforms of CDC25B that are specifically involved in G2/M checkpoint recovery. Cancer Res 71(5):1968-1977.

[45] Kumagai A, Dunphy WG. (1992) Regulation of the cdc25 protein during the cell cycle in Xenopus extracts. Cell 70:139-151.

[46] Rime H, Huchon D, De Smedt V, Thibier C, Galaktionov K, Jessus C, et al. (1994) Microinjection of Cdc25 protein phosphatase into Xenopus prophase oocyte activates MPF and arrests meiosis at metaphase I. Biol Cell 82(1):11-22.

[47] Ashcroft NR, Kosinski ME, Wickramasinghe D, Donovan PJ, Golden A. (1998) The four cdc25 genes from the nematode Caenorhabditis elegans. Gene 214(1-2):59-66.

[48] Alphey L, Jimenez J, White-Cooper H, Dawson I, Nurse P, Glover DM. (1992) twine, a cdc25 homolog that functions in the male and female germline of Drosophila. Cell 69(6):977-988.

[49] Edgar BA, O'Farrell PH. (1989) Genetic control of cell division patterns in the Drosophila embryo. Cell 57(1):177-187.

[50] Kumagai A, Dunphy WG. (1991) The cdc25 protein controls tyrosine dephosphorylation of the cdc2 protein in a cell-free system. Cell 64(5):903-914.

[51] Khadaroo B, Robbens S, Ferraz C, Derelle E, Eychenie S, Cooke R, et al. (2004) The first green lineage cdc25 dual-specificity phosphatase. Cell Cycle 3(4):513-518.

[52] Landrieu I, da Costa M, De Veylder L, Dewitte F, Vandepoele K, Hassan S, et al. (2004) A small CDC25 dual-specificity tyrosine-phosphatase isoform in *Arabidopsis thaliana*. Proc Natl Acad Sci U S A 101(36):13380-13385.

[53] Sorrell DA, Chrimes D, Dickinson JR, Rogers HJ, Francis D. (2005) The Arabidopsis CDC25 induces a short cell length when overexpressed in fission yeast: evidence for cell cycle function. New Phytol 165(2):425-428.

[54] Booher RN, Alfa CE, Hyams JS, Beach DH. (1989) The fission yeast cdc2/cdc13/suc1 protein kinase: regulation of catalytic activity and nuclear localization. Cell 58(3):485-497.

[55] Booher R, Beach D. (1987) Interaction between cdc13+ and cdc2+ in the control of mitosis in fission yeast; dissociation of the G1 and G2 roles of the cdc2+ protein kinase. EMBO J 6(11):3441-3447.

[56] Martin-Castellanos C, Blanco MA, de Prada JM, Moreno S. (2000) The puc1 cyclin regulates the G1 phase of the fission yeast cell cycle in response to cell size. Mol Biol Cell 11(2):543-554.

[57] Durkacz B, Carr A, Nurse P. (1986) Transcription of the cdc2 cell cycle control gene of the fission yeast *Schizosaccharomyces pombe*. EMBO J 5(2):369-373.

[58] Creanor J, Mitchison JM. (1996) The kinetics of the B cyclin p56cdc13 and the phosphatase p80cdc25 during the cell cycle of the fission yeast *Schizosaccharomyces pombe* . J Cell Sci 109 (Pt 6)(Pt 6):1647-1653.

[59] Ducommun B, Draetta G, Young P, Beach D. (1990) Fission yeast cdc25 is a cell-cycle regulated protein. Biochem Biophys Res Commun 167(1):301-309.

[60] Frazer C, Young PG. (2011) Redundant mechanisms prevent mitotic entry following replication arrest in the absence of Cdc25 hyper-phosphorylation in fission yeast. PLoS One 6(6):e21348.

[61] Den Haese GJ, Walworth N, Carr AM, Gould KL. (1995) The Wee1 protein kinase regulates T14 phosphorylation of fission yeast Cdc2. Mol Biol Cell 6(4):371-385.

[62] Gould KL, Nurse P. (1989) Tyrosine phosphorylation of the fission yeast cdc2+ protein kinase regulates entry into mitosis. Nature 342(6245):39-45.

[63] Lundgren K, Walworth N, Booher R, Dembski M, Kirschner M, Beach D. (1991) Mik1 and Wee1 Cooperate in the Inhibitory Tyrosine Phosphorylation of Cdc2. Cell 64(6):1111-1122.

[64] Lee MS, Enoch T, Piwnica-Worms H. (1994) Mik1+ Encodes a Tyrosine Kinase that Phosphorylates P34cdc2 on Tyrosine 15. J Biol Chem 269(48):30530-30537.

[65] Russell P, Nurse P. (1987) Negative regulation of mitosis by wee1+, a gene encoding a protein kinase homolog. Cell 49(4):559-567.

[66] Parker LL, Atherton-Fessler S, Piwnica-Worms H. (1992) P107wee1 is a Dual-Specificity Kinase that Phosphorylates P34cdc2 on Tyrosine 15. Proc Natl Acad Sci U S A 89(7):2917-2921.

[67] Featherstone C, Russell P. (1991) Fission yeast p107wee1 mitotic inhibitor is a tyrosine/serine kinase. Nature 349:808-811.

[68] Chua G, Lingner C, Frazer C, Young PG. (2002) The sal3(+) gene encodes an importin-beta implicated in the nuclear import of Cdc25 in *Schizosaccharomyces pombe*. Genetics 162(2):689-703.

[69] Rupes I, Webb BA, Mak A, Young PG. (2001) G2/M arrest caused by actin disruption is a manifestation of the cell size checkpoint in fission yeast. Mol Biol Cell 12(12):3892-3903.

[70] Kovelman R, Russell P. (1996) Stockpiling of Cdc25 during a DNA replication checkpoint arrest in *Schizosaccharomyces pombe*. Mol Cell Biol 16(1):86-93.

[71] Moreno S, Nurse P. (1990) Substrates for p34cdc2: in vivo veritas? Cell 61(4):549-551.

[72] Shenoy S, Choi JK, Bagrodia S, Copeland TD, Maller JL, Shalloway D. (1989) Purified maturation promoting factor phosphorylates pp60c-src at the sites phosphorylated during fibroblast mitosis. Cell 57(5):763-774.

[73] Wolfe BA, Gould KL. (2004) Fission yeast Clp1p phosphatase affects G2/M transition and mitotic exit through Cdc25p inactivation. EMBO J 23(4):919-929.

[74] Mulvihill DP, Petersen J, Ohkura H, Glover DM, Hagan IM. (1999) Plo1 kinase recruitment to the spindle pole body and its role in cell division in *Schizosaccharomyces pombe*. Mol Biol Cell 10(8):2771-2785.

[75] Charles JF, Jaspersen SL, Tinker-Kulberg RL, Hwang L, Szidon A, Morgan DO. (1998) The Polo-related kinase Cdc5 activates and is destroyed by the mitotic cyclin destruction machinery in S. cerevisiae. Curr Biol 8(9):497-507.

[76] Katsuno Y, Suzuki A, Sugimura K, Okumura K, Zineldeen DH, Shimada M, et al. (2009) Cyclin A-Cdk1 regulates the origin firing program in mammalian cells. Proc Natl Acad Sci U S A 106(9):3184-3189.

[77] Elledge SJ, Spottswood MR. (1991) A new human p34 protein kinase, CDK2, identified by complementation of a cdc28 mutation in *Saccharomyces cerevisiae*, is a homolog of Xenopus Eg1. EMBO J 10(9):2653-2659.

[78] Gu Y, Rosenblatt J, Morgan DO. (1992) Cell cycle regulation of CDK2 activity by phosphorylation of Thr160 and Tyr15. EMBO J 11(11):3995-4005.

[79] Elledge SJ, Richman R, Hall FL, Williams RT, Lodgson N, Harper JW. (1992) CDK2 encodes a 33-kDa cyclin A-associated protein kinase and is expressed before CDC2 in the cell cycle. Proc Natl Acad Sci U S A 89(7):2907-2911.

[80] Rosenblatt J, Gu Y, Morgan DO. (1992) Human cyclin-dependent kinase 2 is activated during the S and G2 phases of the cell cycle and associates with cyclin A. Proc Natl Acad Sci U S A 89(7):2824-2828.

[81] Guadagno TM, Newport JW. (1996) Cdk2 kinase is required for entry into mitosis as a positive regulator of Cdc2-cyclin B kinase activity. Cell 84(1):73-82.

[82] Bates S, Bonetta L, MacAllan D, Parry D, Holder A, Dickson C, et al. (1994) CDK6 (PLSTIRE) and CDK4 (PSK-J3) are a distinct subset of the cyclin-dependent kinases that associate with cyclin D1. Oncogene 9(1):71-79.

[83] Fisher RP, Morgan DO. (1994) A novel cyclin associates with MO15/CDK7 to form the CDK-activating kinase. Cell 78(4):713-724.

[84] Weinberg RA. (1995) The retinoblastoma protein and cell cycle control. Cell 81(3):323-330.

[85] Dyson N. (1998) The regulation of E2F by pRB-family proteins. Genes Dev 12(15):2245-2262.

[86] Vidal A, Koff A. (2000) Cell-cycle inhibitors: three families united by a common cause. Gene 247(1-2):1-15.

[87] Chen X, Prywes R. (1999) Serum-induced expression of the cdc25A gene by relief of E2F-mediated repression. Mol Cell Biol 19(7):4695-4702.

[88] Malumbres M, Pellicer A. (1998) RAS pathways to cell cycle control and cell transformation. Front Biosci 3:d887-912.

[89] Sherr CJ, Roberts JM. (1999) CDK inhibitors: positive and negative regulators of G1-phase progression. Genes Dev 13(12):1501-1512.

[90] Lundberg AS, Weinberg RA. (1998) Functional inactivation of the retinoblastoma protein requires sequential modification by at least two distinct cyclin-cdk complexes. Mol Cell Biol 18(2):753-761.

[91] Blomberg I, Hoffmann I. (1999) Ectopic expression of Cdc25A accelerates the G(1)/S transition and leads to premature activation of cyclin E- and cyclin A-dependent kinases. Mol Cell Biol 19(9):6183-6194.

[92] Sherr CJ, Roberts JM. (2004) Living with or without cyclins and cyclin-dependent kinases. Genes Dev 18(22):2699-2711.

[93] Booher RN, Holman PS, Fattaey A. (1997) Human Myt1 is a cell cycle-regulated kinase that inhibits Cdc2 but not Cdk2 activity. J Biol Chem 272(35):22300-22306.

[94] Sebastian B, Kakizuka A, Hunter T. (1993) Cdc25M2 activation of cyclin-dependent kinases by dephosphorylation of threonine-14 and tyrosine-15. Proc Natl Acad Sci U S A 90(8):3521-3524.

[95] Hoffmann I, Draetta G, Karsenti E. (1994) Activation of the phosphatase activity of human cdc25A by a cdk2-cyclin E dependent phosphorylation at the G1/S transition. EMBO J 13(18):4302-4310.

[96] Gabrielli BG, Clark JM, McCormack AK, Ellem KA. (1997) Hyperphosphorylation of the N-terminal domain of Cdc25 regulates activity toward cyclin B1/Cdc2 but not cyclin A/Cdk2. J Biol Chem 272(45):28607-28614.

[97] Lammer C, Wagerer S, Saffrich R, Mertens D, Ansorge W, Hoffmann I. (1998) The cdc25B phosphatase is essential for the G2/M phase transition in human cells. J Cell Sci 111 (Pt 16)(Pt 16):2445-2453.

[98] Sexl V, Diehl JA, Sherr CJ, Ashmun R, Beach D, Roussel MF. (1999) A rate limiting function of cdc25A for S phase entry inversely correlates with tyrosine dephosphorylation of Cdk2. Oncogene 18(3):573-582.

[99] Ducruet AP, Lazo JS. (2003) Regulation of Cdc25A half-life in interphase by cyclin-dependent kinase 2 activity. J Biol Chem 278(34):31838-31842.

[100] Pagano M, Pepperkok R, Verde F, Ansorge W, Draetta G. (1992) Cyclin A is required at two points in the human cell cycle. EMBO J 11(3):961-971.

[101] Petersen BO, Lukas J, Sorensen CS, Bartek J, Helin K. (1999) Phosphorylation of mammalian CDC6 by cyclin A/CDK2 regulates its subcellular localization. EMBO J 18(2):396-410.

[102] Coverley D, Pelizon C, Trewick S, Laskey RA. (2000) Chromatin-bound Cdc6 persists in S and G2 phases in human cells, while soluble Cdc6 is destroyed in a cyclin A-cdk2 dependent process. J Cell Sci 113 (Pt 11)(Pt 11):1929-1938.

[103] Alexandrow MG, Hamlin JL. (2005) Chromatin decondensation in S-phase involves recruitment of Cdk2 by Cdc45 and histone H1 phosphorylation. J Cell Biol 168(6):875-886.

[104] Garner-Hamrick PA, Fisher C. (1998) Antisense phosphorothioate oligonucleotides specifically down-regulate cdc25B causing S-phase delay and persistent antiproliferative effects. Int J Cancer 76(5):720-728.

[105] Lindqvist A, Kallstrom H, Lundgren A, Barsoum E, Rosenthal CK. (2005) Cdc25B cooperates with Cdc25A to induce mitosis but has a unique role in activating cyclin B1-Cdk1 at the centrosome. J Cell Biol 171(1):35-45.

[106] Chen MS, Hurov J, White LS, Woodford-Thomas T, Piwnica-Worms H. (2001) Absence of apparent phenotype in mice lacking Cdc25C protein phosphatase. Mol Cell Biol 21(12):3853-3861.

[107] Lincoln AJ, Wickramasinghe D, Stein P, Schultz RM, Palko ME, De Miguel MP, et al. (2002) Cdc25b phosphatase is required for resumption of meiosis during oocyte maturation. Nat Genet 30(4):446-449.

[108] Ferguson AM, White LS, Donovan PJ, Piwnica-Worms H. (2005) Normal cell cycle and checkpoint responses in mice and cells lacking Cdc25B and Cdc25C protein phosphatases. Mol Cell Biol 25(7):2853-2860.

[109] Furuno N, den Elzen N, Pines J. (1999) Human cyclin A is required for mitosis until mid prophase. J Cell Biol 147:295-306.

[110] Mitra J, Enders GH. (2004) Cyclin A/Cdk2 complexes regulate activation of Cdk1 and Cdc25 phosphatases in human cells. Oncogene 23(19):3361-3367.

[111] Minshull J, Golsteyn R, Hill CS, Hunt T. (1990) The A- and B-type cyclin associated cdc2 kinases in Xenopus turn on and off at different times in the cell cycle. EMBO J 9(9):2865-2875.

[112] Cans C, Ducommun B, Baldin V. (1999) Proteasome-dependent degradation of human CDC25B phosphatase. Mol Biol Rep 26(1-2):53-57.

[113] Baldin V, Cans C, Knibiehler M, Ducommun B. (1997) Phosphorylation of human CDC25B phosphatase by CDK1-cyclin A triggers its proteasome-dependent degradation. J Biol Chem 272(52):32731-32734.

[114] Gabrielli BG, Lee MS, Walker DH, Piwnica-Worms H, Maller JL. (1992) Cdc25 regulates the phosphorylation and activity of the Xenopus cdk2 protein kinase complex. J Biol Chem 267(25):18040-18046.

[115] Malumbres M, Barbacid M. (2005) Mammalian cyclin-dependent kinases. Trends Biochem Sci 30(11):630-641.

[116] De Boer L, Oakes V, Beamish H, Giles N, Stevens F, Somodevilla-Torres M, et al. (2008) Cyclin A/cdk2 coordinates centrosomal and nuclear mitotic events. Oncogene 27(31):4261-4268.

[117] Nishijima H, Nishitani H, Seki T, Nishimoto T. (1997) A dual-specificity phosphatase Cdc25B is an unstable protein and triggers p34(cdc2)/cyclin B activation in hamster BHK21 cells arrested with hydroxyurea. J Cell Biol 138(5):1105-1116.

[118] Timofeev O, Cizmecioglu O, Settele F, Kempf T, Hoffmann I. (2010) Cdc25 phosphatases are required for timely assembly of CDK1-cyclin B at the G2/M transition. J Biol Chem 285(22):16978-16990.

[119] Bailly E, Doree M, Nurse P, Bornens M. (1989) p34cdc2 is located in both nucleus and cytoplasm; part is centrosomally associated at G2/M and enters vesicles at anaphase. EMBO J 8(13):3985-3995.

[120] Gabrielli BG, De Souza CP, Tonks ID, Clark JM, Hayward NK, Ellem KA. (1996) Cytoplasmic accumulation of cdc25B phosphatase in mitosis triggers centrosomal microtubule nucleation in HeLa cells. J Cell Sci 109 (Pt 5)(Pt 5):1081-1093.

[121] Pines J, Hunter T. (1991) Human cyclins A and B1 are differentially located in the cell and undergo cell cycle-dependent nuclear transport. J Cell Biol 115(1):1-17.

[122] Hirota T, Kunitoku N, Sasayama T, Marumoto T, Zhang D, Nitta M, et al. (2003) Aurora-A and an interacting activator, the LIM protein Ajuba, are required for mitotic commitment in human cells. Cell 114(5):585-598.

[123] Barr AR, Gergely F. (2007) Aurora-A: the maker and breaker of spindle poles. J Cell Sci 120(Pt 17):2987-2996.

[124] Dutertre S, Cazales M, Quaranta M, Froment C, Trabut V, Dozier C, et al. (2004) Phosphorylation of CDC25B by Aurora-A at the centrosome contributes to the G2-M transition. J Cell Sci 117(Pt 12):2523-2531.

[125] Jackman M, Lindon C, Nigg EA, Pines J. (2003) Active cyclin B1-Cdk1 first appears on centrosomes in prophase. Nat Cell Biol 5(2):143-148.

[126] Pines J, Hunter T. (1994) The differential localization of human cyclins A and B is due to a cytoplasmic retention signal in cyclin B. EMBO J 13(16):3772-3781.

[127] Hagting A, Karlsson C, Clute P, Jackman M, Pines J. (1998) MPF localization is controlled by nuclear export. EMBO J 17(14):4127-4138.

[128] Toyoshima F, Moriguchi T, Wada A, Fukuda M, Nishida E. (1998) Nuclear export of cyclin B1 and its possible role in the DNA damage-induced G2 checkpoint. EMBO J 17(10):2728-2735.

[129] Yang J, Bardes ES, Moore JD, Brennan J, Powers MA, Kornbluth S. (1998) Control of cyclin B1 localization through regulated binding of the nuclear export factor CRM1. Genes Dev 12(14):2131-2143.

[130] Borgne A, Ostvold AC, Flament S, Meijer L. (1999) Intra-M phase-promoting factor phosphorylation of cyclin B at the prophase/metaphase transition. J Biol Chem 274(17):11977-11986.

[131] Toyoshima-Morimoto F, Taniguchi E, Shinya N, Iwamatsu A, Nishida E. (2001) Polo-like kinase 1 phosphorylates cyclin B1 and targets it to the nucleus during prophase. Nature 410(6825):215-220.

[132] Okano-Uchida T, Okumura E, Iwashita M, Yoshida H, Tachibana K, Kishimoto T. (2003) Distinct regulators for Plk1 activation in starfish meiotic and early embryonic cycles. EMBO J 22(20):5633-5642.

[133] Nakajima H, Toyoshima-Morimoto F, Taniguchi E, Nishida E. (2003) Identification of a consensus motif for Plk (Polo-like kinase) phosphorylation reveals Myt1 as a Plk1 substrate. J Biol Chem 278(28):25277-25280.

[134] Watanabe N, Arai H, Nishihara Y, Taniguchi M, Watanabe N, Hunter T, et al. (2004) M-phase kinases induce phospho-dependent ubiquitination of somatic Wee1 by SCFbeta-TrCP. Proc Natl Acad Sci U S A 101(13):4419-4424.

[135] Baldin V, Pelpel K, Cazales M, Cans C, Ducommun B. (2002) Nuclear localization of CDC25B1 and serine 146 integrity are required for induction of mitosis. J Biol Chem 277(38):35176-35182.

[136] Lobjois V, Jullien D, Bouche JP, Ducommun B. (2009) The polo-like kinase 1 regulates CDC25B-dependent mitosis entry. Biochim Biophys Acta 1793(3):462-468.

[137] Lobjois V, Froment C, Braud E, Grimal F, Burlet-Schiltz O, Ducommun B, et al. (2011) Study of the docking-dependent PLK1 phosphorylation of the CDC25B phosphatase. Biochem Biophys Res Commun 410(1):87-90.

[138] Qian YW, Erikson E, Taieb FE, Maller JL. (2001) The polo-like kinase Plx1 is required for activation of the phosphatase Cdc25C and cyclin B-Cdc2 in Xenopus oocytes. Mol Biol Cell 12(6):1791-1799.

[139] Roshak AK, Capper EA, Imburgia C, Fornwald J, Scott G, Marshall LA. (2000) The human polo-like kinase, PLK, regulates cdc2/cyclin B through phosphorylation and activation of the cdc25C phosphatase. Cell Signal 12(6):405-411.

[140] Toyoshima-Morimoto F, Taniguchi E, Nishida E. (2002) Plk1 promotes nuclear translocation of human Cdc25C during prophase. EMBO Rep 3(4):341-348.

[141] Bahassi el M, Hennigan RF, Myer DL, Stambrook PJ. (2004) Cdc25C phosphorylation on serine 191 by Plk3 promotes its nuclear translocation. Oncogene 23(15):2658-2663.

[142] Bonni S, Ganuelas ML, Petrinac S, Hudson JW. (2008) Human Plk4 phosphorylates Cdc25C. Cell Cycle 7(4):545-547.

[143] Strausfeld U, Fernandez A, Capony JP, Girard F, Lautredou N, Derancourt J, et al. (1994) Activation of p34cdc2 protein kinase by microinjection of human cdc25C into mammalian cells. Requirement for prior phosphorylation of cdc25C by p34cdc2 on sites phosphorylated at mitosis. J Biol Chem 269(8):5989-6000.

[144] Honda R, Ohba Y, Nagata A, Okayama H, Yasuda H. (1993) Dephosphorylation of human p34cdc2 kinase on both Thr-14 and Tyr-15 by human cdc25B phosphatase. FEBS Lett 318(3):331-334.

[145] Franckhauser C, Mamaeva D, Heron-Milhavet L, Fernandez A, Lamb NJ. (2010) Distinct pools of cdc25C are phosphorylated on specific TP sites and differentially localized in human mitotic cells. PLoS One 5(7):e11798.

[146] Izumi T, Maller JL. (1993) Elimination of cdc2 phosphorylation sites in the cdc25 phosphatase blocks initiation of M-phase. Mol Biol Cell 4(12):1337-1350.

[147] Pines J, Hunter T. Cyclins A and B1 in the Human Cell Cycle. In: Marsh J, editor. Regulation of the Eukaryotic Cell New York: Wiley; 1992. p. 187-208.

[148] Mailand N, Podtelejnikov AV, Groth A, Mann M, Bartek J, Lukas J. (2002) Regulation of G(2)/M events by Cdc25A through phosphorylation-dependent modulation of its stability. EMBO J 21(21):5911-5920.

[149] Tumurbaatar I, Cizmecioglu O, Hoffmann I, Grummt I, Voit R. (2011) Human Cdc14B promotes progression through mitosis by dephosphorylating Cdc25 and regulating Cdk1/cyclin B activity. PLoS One 6(2):e14711.

[150] Barford D. (2011) Structural insights into anaphase-promoting complex function and mechanism. Philos Trans R Soc Lond B Biol Sci 366(1584):3605-3624.

[151] Esteban V, Vazquez-Novelle MD, Calvo E, Bueno A, Sacristan MP. (2006) Human Cdc14A reverses CDK1 phosphorylation of Cdc25A on serines 115 and 320. Cell Cycle 5(24):2894-2898.

[152] Donzelli M, Squatrito M, Ganoth D, Hershko A, Pagano M, Draetta GF. (2002) Dual mode of degradation of Cdc25 A phosphatase. EMBO J 21(18):4875-4884.

[153] Kieffer I, Lorenzo C, Dozier C, Schmitt E, Ducommun B. (2007) Differential mitotic degradation of the CDC25B phosphatase variants. Oncogene 26(57):7847-7858.

[154] Rhind N, Furnari B, Russell P. (1997) Cdc2 tyrosine phosphorylation is required for the DNA damage checkpoint in fission yeast. Genes Dev 11(4):504-511.

[155] Boddy MN, Furnari B, Mondesert O, Russell P. (1998) Replication checkpoint enforced by kinases Cds1 and Chk1. Science 280(5365):909-912.

[156] Zeng Y, Forbes KC, Wu Z, Moreno S, Piwnica-Worms H, Enoch T. (1998) Replication checkpoint requires phosphorylation of the phosphatase Cdc25 by Cds1 or Chk1. Nature 395(6701):507-510.

[157] Furnari B, Blasina A, Boddy MN, McGowan CH, Russell P. (1999) Cdc25 inhibited in vivo and in vitro by checkpoint kinases Cds1 and Chk1. Mol Biol Cell 10(4):833-845.

[158] Branzei D, Foiani M. (2010) Maintaining genome stability at the replication fork. Nat Rev Mol Cell Biol 11(3):208-219.

[159] Melo J, Toczyski D. (2002) A unified view of the DNA-damage checkpoint. Curr Opin Cell Biol 14(2):237-245.

[160] Lopez-Girona A, Furnari B, Mondesert O, Russell P. (1999) Nuclear localization of Cdc25 is regulated by DNA damage and a 14-3-3 protein. Nature 397(6715):172-175.

[161] Ford JC, al-Khodairy F, Fotou E, Sheldrick KS, Griffiths DJ, Carr AM. (1994) 14-3-3 protein homologs required for the DNA damage checkpoint in fission yeast. Science 265(5171):533-535.

[162] Zeng Y, Piwnica-Worms H. (1999) DNA damage and replication checkpoints in fission yeast require nuclear exclusion of the Cdc25 phosphatase via 14-3-3 binding. Mol Cell Biol 19(11):7410-7419.

[163] Dai Y, Grant S. (2010) New insights into checkpoint kinase 1 in the DNA damage response signaling network. Clin Cancer Res 16(2):376-383.

[164] Niida H, Nakanishi M. (2006) DNA damage checkpoints in mammals. Mutagenesis 21(1):3-9.

[165] Minella AC, Grim JE, Welcker M, Clurman BE. (2007) p53 and SCFFbw7 cooperatively restrain cyclin E-associated genome instability. Oncogene 26(48):6948-6953.

[166] Terada Y, Tatsuka M, Jinno S, Okayama H. (1995) Requirement for tyrosine phosphorylation of Cdk4 in G1 arrest induced by ultraviolet irradiation. Nature 376(6538):358-362.

[167] Mailand N, Falck J, Lukas C, Syljuasen RG, Welcker M, Bartek J, et al. (2000) Rapid destruction of human Cdc25A in response to DNA damage. Science 288(5470):1425-1429.

[168] Goloudina A, Yamaguchi H, Chervyakova DB, Appella E, Fornace AJ,Jr, Bulavin DV. (2003) Regulation of human Cdc25A stability by Serine 75 phosphorylation is not sufficient to activate a S phase checkpoint. Cell Cycle 2(5):473-478.

[169] Hassepass I, Voit R, Hoffmann I. (2003) Phosphorylation at serine 75 is required for UV-mediated degradation of human Cdc25A phosphatase at the S-phase checkpoint. J Biol Chem 278(32):29824-29829.

[170] Chen MS, Ryan CE, Piwnica-Worms H. (2003) Chk1 kinase negatively regulates mitotic function of Cdc25A phosphatase through 14-3-3 binding. Mol Cell Biol 23(21):7488-7497.

[171] Jin J, Shirogane T, Xu L, Nalepa G, Qin J, Elledge SJ, et al. (2003) SCFbeta-TRCP links Chk1 signaling to degradation of the Cdc25A protein phosphatase. Genes Dev 17(24):3062-3074.

[172] Zhao H, Watkins JL, Piwnica-Worms H. (2002) Disruption of the checkpoint kinase 1/cell division cycle 25A pathway abrogates ionizing radiation-induced S and G2 checkpoints. Proc Natl Acad Sci U S A 99(23):14795-14800.

[173] Falck J, Mailand N, Syljuasen RG, Bartek J, Lukas J. (2001) The ATM-Chk2-Cdc25A checkpoint pathway guards against radioresistant DNA synthesis. Nature 410(6830):842-847.

[174] Shreeram S, Hee WK, Bulavin DV. (2008) Cdc25A serine 123 phosphorylation couples centrosome duplication with DNA replication and regulates tumorigenesis. Mol Cell Biol 28(24):7442-7450.

[175] Madlener S, Rosner M, Krieger S, Giessrigl B, Gridling M, Vo TP, et al. (2009) Short 42 degrees C heat shock induces phosphorylation and degradation of Cdc25A which depends on p38MAPK, Chk2 and 14.3.3. Hum Mol Genet 18(11):1990-2000.

[176] Falck J, Petrini JH, Williams BR, Lukas J, Bartek J. (2002) The DNA damage-dependent intra-S phase checkpoint is regulated by parallel pathways. Nat Genet 30(3):290-294.

[177] Yaffe MB, Rittinger K, Volinia S, Caron PR, Aitken A, Leffers H, et al. (1997) The structural basis for 14-3-3:phosphopeptide binding specificity. Cell 91(7):961-971.

[178] Kasahara K, Goto H, Enomoto M, Tomono Y, Kiyono T, Inagaki M. (2010) 14-3-3gamma mediates Cdc25A proteolysis to block premature mitotic entry after DNA damage. EMBO J 29(16):2802-2812.

[179] Aitken A. (2002) Functional specificity in 14-3-3 isoform interactions through dimer formation and phosphorylation. Chromosome location of mammalian isoforms and variants. Plant Mol Biol 50(6):993-1010.

[180] Aitken A, Baxter H, Dubois T, Clokie S, Mackie S, Mitchell K, et al. (2002) Specificity of 14-3-3 isoform dimer interactions and phosphorylation. Biochem Soc Trans 30:351-360.

[181] Xiao Z, Chen Z, Gunasekera AH, Sowin TJ, Rosenberg SH, Fesik S, et al. (2003) Chk1 mediates S and G2 arrests through Cdc25A degradation in response to DNA-damaging agents. J Biol Chem 278(24):21767-21773.

[182] Costanzo V, Robertson K, Ying CY, Kim E, Avvedimento E, Gottesman M, et al. (2000) Reconstitution of an ATM-dependent checkpoint that inhibits chromosomal DNA replication following DNA damage. Mol Cell 6(3):649-659.

[183] Molinari M, Mercurio C, Dominguez J, Goubin F, Draetta GF. (2000) Human Cdc25 A inactivation in response to S phase inhibition and its role in preventing premature mitosis. EMBO Rep 1(1):71-79.
[184] Busino L, Donzelli M, Chiesa M, Guardavaccaro D, Ganoth D, Dorrello NV, et al. (2003) Degradation of Cdc25A by beta-TrCP during S phase and in response to DNA damage. Nature 426(6962):87-91.
[185] Winston JT, Strack P, Beer-Romero P, Chu CY, Elledge SJ, Harper JW. (1999) The SCFbeta-TRCP-ubiquitin ligase complex associates specifically with phosphorylated destruction motifs in IkappaBalpha and beta-catenin and stimulates IkappaBalpha ubiquitination in vitro. Genes Dev 13(3):270-283.
[186] Donzelli M, Busino L, Chiesa M, Ganoth D, Hershko A, Draetta GF. (2004) Hierarchical order of phosphorylation events commits Cdc25A to betaTrCP-dependent degradation. Cell Cycle 3(4):469-471.
[187] Melixetian M, Klein DK, Sorensen CS, Helin K. (2009) NEK11 regulates CDC25A degradation and the IR-induced G2/M checkpoint. Nat Cell Biol 11(10):1247-1253.
[188] Shieh SY, Ahn J, Tamai K, Taya Y, Prives C. (2000) The human homologs of checkpoint kinases Chk1 and Cds1 (Chk2) phosphorylate p53 at multiple DNA damage-inducible sites. Genes Dev 14(3):289-300.
[189] Khosravi R, Maya R, Gottlieb T, Oren M, Shiloh Y, Shkedy D. (1999) Rapid ATM-dependent phosphorylation of MDM2 precedes p53 accumulation in response to DNA damage. Proc Natl Acad Sci U S A 96(26):14973-14977.
[190] Maya R, Balass M, Kim ST, Shkedy D, Leal JF, Shifman O, et al. (2001) ATM-dependent phosphorylation of Mdm2 on serine 395: role in p53 activation by DNA damage. Genes Dev 15(9):1067-1077.
[191] Tritarelli A, Oricchio E, Ciciarello M, Mangiacasale R, Palena A, Lavia P, et al. (2004) p53 localization at centrosomes during mitosis and postmitotic checkpoint are ATM-dependent and require serine 15 phosphorylation. Mol Biol Cell 15:3751.
[192] el-Deiry WS, Harper JW, O'Connor PM, Velculescu VE, Canman CE, Jackman J, et al. (1994) WAF1/CIP1 is induced in p53-mediated G1 arrest and apoptosis. Cancer Res 54(5):1169-1174.
[193] Watcharasit P, Bijur GN, Zmijewski JW, Song L, Zmijewska A, Chen X, et al. (2002) Direct, activating interaction between glycogen synthase kinase-3beta and p53 after DNA damage. Proc Natl Acad Sci U S A 99(12):7951-7955.
[194] Kang T, Wei Y, Honaker Y, Yamaguchi H, Appella E, Hung MC, et al. (2008) GSK-3 beta targets Cdc25A for ubiquitin-mediated proteolysis, and GSK-3 beta inactivation correlates with Cdc25A overproduction in human cancers. Cancer Cell 13(1):36-47.
[195] Myer DL, Robbins SB, Yin M, Boivin GP, Liu Y, Greis KD, et al. (2011) Absence of polo-like kinase 3 in mice stabilizes Cdc25A after DNA damage but is not sufficient to produce tumors. Mutat Res 714(1-2):1-10.
[196] Xie S, Wu H, Wang Q, Cogswell JP, Husain I, Conn C, et al. (2001) Plk3 functionally links DNA damage to cell cycle arrest and apoptosis at least in part via the p53 pathway. J Biol Chem 276(46):43305-43312.
[197] Bahassi el M, Myer DL, McKenney RJ, Hennigan RF, Stambrook PJ. (2006) Priming phosphorylation of Chk2 by polo-like kinase 3 (Plk3) mediates its full activation by

ATM and a downstream checkpoint in response to DNA damage. Mutat Res 596(1-2):166-176.

[198] Honaker Y, Piwnica-Worms H. (2010) Casein kinase 1 functions as both penultimate and ultimate kinase in regulating Cdc25A destruction. Oncogene 29(23):3324-3334.

[199] Piao S, Lee SJ, Xu Y, Gwak J, Oh S, Park BJ, et al. (2011) CK1epsilon targets Cdc25A for ubiquitin-mediated proteolysis under normal conditions and in response to checkpoint activation. Cell Cycle 10(3):531-537.

[200] Sorensen CS, Syljuasen RG, Falck J, Schroeder T, Ronnstrand L, Khanna KK, et al. (2003) Chk1 regulates the S phase checkpoint by coupling the physiological turnover and ionizing radiation-induced accelerated proteolysis of Cdc25A. Cancer Cell 3(3):247-258.

[201] Shimuta K, Nakajo N, Uto K, Hayano Y, Okazaki K, Sagata N. (2002) Chk1 is activated transiently and targets Cdc25A for degradation at the Xenopus midblastula transition. EMBO J 21(14):3694-3703.

[202] Kanemori Y, Uto K, Sagata N. (2005) Beta-TrCP recognizes a previously undescribed nonphosphorylated destruction motif in Cdc25A and Cdc25B phosphatases. Proc Natl Acad Sci U S A 102(18):6279-6284.

[203] Koledova Z, Kafkova LR, Kramer A, Divoky V. (2010) DNA damage-induced degradation of Cdc25A does not lead to inhibition of Cdk2 activity in mouse embryonic stem cells. Stem Cells 28(3):450-461.

[204] Thomas Y, Coux O, Baldin V. (2010) betaTrCP-dependent degradation of CDC25B phosphatase at the metaphase-anaphase transition is a pre-requisite for correct mitotic exit. Cell Cycle 9(21):4338-4350.

[205] Bansal P, Lazo JS. (2007) Induction of Cdc25B regulates cell cycle resumption after genotoxic stress. Cancer Res 67(7):3356-3363.

[206] Jin J, Ang XL, Ye X, Livingstone M, Harper JW. (2008) Differential roles for checkpoint kinases in DNA damage-dependent degradation of the Cdc25A protein phosphatase. J Biol Chem 283(28):19322-19328.

[207] Uto K, Inoue D, Shimuta K, Nakajo N, Sagata N. (2004) Chk1, but not Chk2, inhibits Cdc25 phosphatases by a novel common mechanism. EMBO J 23(16):3386-3396.

[208] Bulavin DV, Higashimoto Y, Popoff IJ, Gaarde WA, Basrur V, Potapova O, et al. (2001) Initiation of a G2/M checkpoint after ultraviolet radiation requires p38 kinase. Nature 411(6833):102-107.

[209] Matsuoka S, Huang M, Elledge SJ. (1998) Linkage of ATM to cell cycle regulation by the Chk2 protein kinase. Science 282(5395):1893-1897.

[210] Manke IA, Nguyen A, Lim D, Stewart MQ, Elia AE, Yaffe MB. (2005) MAPKAP kinase-2 is a cell cycle checkpoint kinase that regulates the G2/M transition and S phase progression in response to UV irradiation. Mol Cell 17(1):37-48.

[211] Uchida S, Yoshioka K, Kizu R, Nakagama H, Matsunaga T, Ishizaka Y, et al. (2009) Stress-activated mitogen-activated protein kinases c-Jun NH2-terminal kinase and p38 target Cdc25B for degradation. Cancer Res 69(16):6438-6444.

[212] Galaktionov K, Lee AK, Eckstein J, Draetta G, Meckler J, Loda M, et al. (1995) CDC25 phosphatases as potential human oncogenes. Science 269(5230):1575-1577.

[213] Gabrielli BG, Clark JM, McCormack AK, Ellem KA. (1997) Ultraviolet light-induced G2 phase cell cycle checkpoint blocks cdc25-dependent progression into mitosis. Oncogene 15(7):749-758.

[214] Lemaire M, Ducommun B, Nebreda AR. (2010) UV-induced downregulation of the CDC25B protein in human cells. FEBS Lett 584(6):1199-1204.

[215] Deng J, Harding HP, Raught B, Gingras AC, Berlanga JJ, Scheuner D, et al. (2002) Activation of GCN2 in UV-irradiated cells inhibits translation. Curr Biol 12(15):1279-1286.

[216] Uchida S, Kuma A, Ohtsubo M, Shimura M, Hirata M, Nakagama H, et al. (2004) Binding of 14-3-3beta but not 14-3-3sigma controls the cytoplasmic localization of CDC25B: binding site preferences of 14-3-3 subtypes and the subcellular localization of CDC25B. J Cell Sci 117(Pt 14):3011-3020.

[217] Dalal SN, Schweitzer CM, Gan J, DeCaprio JA. (1999) Cytoplasmic localization of human cdc25C during interphase requires an intact 14-3-3 binding site. Mol Cell Biol 19(6):4465-4479.

[218] Barth H, Hoffmann I, Kinzel V. (1996) Radiation with 1 Gy prevents the activation of the mitotic inducers mitosis-promoting factor (MPF) and cdc25-C in HeLa cells. Cancer Res 56(10):2268-2272.

[219] Peng CY, Graves PR, Thoma RS, Wu Z, Shaw AS, Piwnica-Worms H. (1997) Mitotic and G2 checkpoint control: regulation of 14-3-3 protein binding by phosphorylation of Cdc25C on serine-216. Science 277(5331):1501-1505.

[220] Sanchez Y, Wong C, Thoma RS, Richman R, Wu Z, Piwnica-Worms H, et al. (1997) Conservation of the Chk1 checkpoint pathway in mammals: linkage of DNA damage to Cdk regulation through Cdc25. Science 277(5331):1497-1501.

[221] Ogg S, Gabrielli B, Piwnica-Worms H. (1994) Purification of a serine kinase that associates with and phosphorylates human Cdc25C on serine 216. J Biol Chem 269(48):30461-30469.

[222] Brown AL, Lee CH, Schwarz JK, Mitiku N, Piwnica-Worms H, Chung JH. (1999) A human Cds1-related kinase that functions downstream of ATM protein in the cellular response to DNA damage. Proc Natl Acad Sci U S A 96(7):3745-3750.

[223] Morris MC, Heitz A, Mery J, Heitz F, Divita G. (2000) An essential phosphorylation-site domain of human cdc25C interacts with both 14-3-3 and cyclins. J Biol Chem 275(37):28849-28857.

[224] Graves PR, Lovly CM, Uy GL, Piwnica-Worms H. (2001) Localization of human Cdc25C is regulated both by nuclear export and 14-3-3 protein binding. Oncogene 20(15):1839-1851.

[225] Chan PM, Ng YW, Manser E. (2011) A robust protocol to map binding sites of the 14-3-3 interactome: Cdc25C requires phosphorylation of both S216 and S263 to bind 14-3-3. Mol Cell Proteomics 10(3):M110.005157.

[226] Esmenjaud-Mailhat C, Lobjois V, Froment C, Golsteyn RM, Monsarrat B, Ducommun B. (2007) Phosphorylation of CDC25C at S263 controls its intracellular localisation. FEBS Lett 581(21):3979-3985.

[227] O'Neill T, Giarratani L, Chen P, Iyer L, Lee CH, Bobiak M, et al. (2002) Determination of substrate motifs for human Chk1 and hCds1/Chk2 by the oriented peptide library approach. J Biol Chem 277(18):16102-16115.

[228] Kumagai A, Dunphy WG. (1999) Binding of 14-3-3 proteins and nuclear export control the intracellular localization of the mitotic inducer Cdc25. Genes Dev 13(9):1067-1072.

[229] Kumagai A, Yakowec PS, Dunphy WG. (1998) 14-3-3 proteins act as negative regulators of the mitotic inducer Cdc25 in Xenopus egg extracts. Mol Biol Cell 9(2):345-354.

[230] Kumagai A, Guo Z, Emami KH, Wang SX, Dunphy WG. (1998) The Xenopus Chk1 protein kinase mediates a caffeine-sensitive pathway of checkpoint control in cell-free extracts. J Cell Biol 142(6):1559-1569.

[231] Hutchins JR, Dikovskaya D, Clarke PR. (2002) Dephosphorylation of the inhibitory phosphorylation site S287 in Xenopus Cdc25C by protein phosphatase-2A is inhibited by 14-3-3 binding. FEBS Lett 528(1-3):267-271.

[232] Sorensen CS, Syljuasen RG, Lukas J, Bartek J. (2004) ATR, Claspin and the Rad9-Rad1-Hus1 complex regulate Chk1 and Cdc25A in the absence of DNA damage. Cell Cycle 3(7):941-945.

[233] Kreutzer J, Guerra B. (2007) The regulatory beta-subunit of protein kinase CK2 accelerates the degradation of CDC25A phosphatase through the checkpoint kinase Chk1. Int J Oncol 31(5):1251-1259.

[234] Schwindling SL, Noll A, Montenarh M, Gotz C. (2004) Mutation of a CK2 phosphorylation site in cdc25C impairs importin alpha/beta binding and results in cytoplasmic retention. Oncogene 23(23):4155-4165.

[235] Theis-Febvre N, Filhol O, Froment C, Cazales M, Cochet C, Monsarrat B, et al. (2003) Protein kinase CK2 regulates CDC25B phosphatase activity. Oncogene 22(2):220-232.

[236] Schmitt E, Boutros R, Froment C, Monsarrat B, Ducommun B, Dozier C. (2006) CHK1 phosphorylates CDC25B during the cell cycle in the absence of DNA damage. J Cell Sci 119(Pt 20):4269-4275.

[237] Kramer A, Mailand N, Lukas C, Syljuasen RG, Wilkinson CJ, Nigg EA, et al. (2004) Centrosome-associated Chk1 prevents premature activation of cyclin-B-Cdk1 kinase. Nat Cell Biol 6(9):884-891.

[238] Forrest A, Gabrielli B. (2001) Cdc25B activity is regulated by 14-3-3. Oncogene 20(32):4393-4401.

[239] Giles N, Forrest A, Gabrielli B. (2003) 14-3-3 acts as an intramolecular bridge to regulate cdc25B localization and activity. J Biol Chem 278(31):28580-28587.

[240] Davezac N, Baldin V, Gabrielli B, Forrest A, Theis-Febvre N, Yashida M, et al. (2000) Regulation of CDC25B phosphatases subcellular localization. Oncogene 19(18):2179-2185.

[241] Stanford JS, Ruderman JV. (2005) Changes in regulatory phosphorylation of Cdc25C Ser287 and Wee1 Ser549 during normal cell cycle progression and checkpoint arrests. Mol Biol Cell 16(12):5749-5760.

[242] Brown EJ, Baltimore D. (2003) Essential and dispensable roles of ATR in cell cycle arrest and genome maintenance. Genes Dev 17(5):615-628.

[243] Liu Q, Guntuku S, Cui XS, Matsuoka S, Cortez D, Tamai K, et al. (2000) Chk1 is an essential kinase that is regulated by Atr and required for the G(2)/M DNA damage checkpoint. Genes Dev 14(12):1448-1459.

[244] Yang J, Winkler K, Yoshida M, Kornbluth S. (1999) Maintenance of G2 arrest in the Xenopus oocyte: a role for 14-3-3-mediated inhibition of Cdc25 nuclear import. EMBO J 18(8):2174-2183.

[245] Brondello JM, Boddy MN, Furnari B, Russell P. (1999) Basis for the checkpoint signal specificity that regulates Chk1 and Cds1 protein kinases. Mol Cell Biol 19(6):4262-4269.

[246] Bulavin DV, Higashimoto Y, Demidenko ZN, Meek S, Graves P, Phillips C, et al. (2003) Dual phosphorylation controls Cdc25 phosphatases and mitotic entry. Nat Cell Biol 5(6):545-551.

[247] Bulavin DV, Demidenko ZN, Phillips C, Moody SA, Fornace AJ,Jr. (2003) Phosphorylation of Xenopus Cdc25C at Ser285 interferes with ability to activate a DNA damage replication checkpoint in pre-midblastula embryos. Cell Cycle 2(3):263-266.

[248] Margolis SS, Perry JA, Forester CM, Nutt LK, Guo Y, Jardim MJ, et al. (2006) Role for the PP2A/B56delta phosphatase in regulating 14-3-3 release from Cdc25 to control mitosis. Cell 127(4):759-773.

[249] Margolis SS, Perry JA, Weitzel DH, Freel CD, Yoshida M, Haystead TA, et al. (2006) A role for PP1 in the Cdc2/Cyclin B-mediated positive feedback activation of Cdc25. Mol Biol Cell 17(4):1779-1789.

[250] Cunat S, Anahory T, Berthenet C, Hedon B, Franckhauser C, Fernandez A, et al. (2008) The cell cycle control protein cdc25C is present, and phosphorylated on serine 214 in the transition from germinal vesicle to metaphase II in human oocyte meiosis. Mol Reprod Dev 75(7):1176-1184.

[251] Margolis SS, Walsh S, Weiser DC, Yoshida M, Shenolikar S, Kornbluth S. (2003) PP1 control of M phase entry exerted through 14-3-3-regulated Cdc25 dephosphorylation. EMBO J 22(21):5734-5745.

[252] Ku NO, Michie S, Resurreccion EZ, Broome RL, Omary MB. (2002) Keratin binding to 14-3-3 proteins modulates keratin filaments and hepatocyte mitotic progression. Proc Natl Acad Sci U S A 99(7):4373-4378.

[253] Clarke PR, Hoffmann I, Draetta G, Karsenti E. (1993) Dephosphorylation of cdc25-C by a type-2A protein phosphatase: specific regulation during the cell cycle in Xenopus egg extracts. Mol Biol Cell 4(4):397-411.

[254] Crenshaw DG, Yang J, Means AR, Kornbluth S. (1998) The mitotic peptidyl-prolyl isomerase, Pin1, interacts with Cdc25 and Plx1. EMBO J 17(5):1315-1327.

[255] Zhou XZ, Kops O, Werner A, Lu PJ, Shen M, Stoller G, et al. (2000) Pin1-dependent prolyl isomerization regulates dephosphorylation of Cdc25C and tau proteins. Mol Cell 6(4):873-883.

[256] Stukenberg PT, Kirschner MW. (2001) Pin1 acts catalytically to promote a conformational change in Cdc25. Mol Cell 7(5):1071-1083.

[257] Winkler KE, Swenson KI, Kornbluth S, Means AR. (2000) Requirement of the prolyl isomerase Pin1 for the replication checkpoint. Science 287(5458):1644-1647.

[258] Forester CM, Maddox J, Louis JV, Goris J, Virshup DM. (2007) Control of mitotic exit by PP2A regulation of Cdc25C and Cdk1. Proc Natl Acad Sci U S A 104(50):19867-19872.

[259] Astuti P, Boutros R, Ducommun B, Gabrielli B. (2010) Mitotic phosphorylation of Cdc25B Ser321 disrupts 14-3-3 binding to the high affinity Ser323 site. J Biol Chem 285(45):34364-34370.

[260] Hermeking H, Lengauer C, Polyak K, He TC, Zhang L, Thiagalingam S, et al. (1997) 14-3-3 sigma is a p53-regulated inhibitor of G2/M progression. Mol Cell 1(1):3-11.

[261] Wilson-Grady JT, Villén J, Gygi SP. (2008) Phosphoproteome analysis of fission yeast. Journal of Proteome Research 7(3):1088-1097.

[262] Kinoshita N, Ohkura H, Yanagida M. (1990) Distinct, essential roles of type 1 and 2A protein phosphatases in the control of the fission yeast cell division cycle. Cell 63(2):405-415.

[263] Kinoshita N, Yamano H, Niwa H, Yoshida T, Yanagida M. (1993) Negative regulation of mitosis by the fission yeast protein phosphatase ppa2. Genes Dev 7(6):1059-1071.

[264] Ohkura H, Adachi Y, Kinoshita N, Niwa O, Toda T, Yanagida M. (1988) Cold-sensitive and caffeine-supersensitive mutants of the Schizosaccharomyces pombe dis genes implicated in sister chromatid separation during mitosis. EMBO J 7(5):1465-1473.

[265] Cargnello M, Roux PP. (2011) Activation and function of the MAPKs and their substrates, the MAPK-activated protein kinases. Microbiol Mol Biol Rev 75(1):50-83.

[266] Shaul YD, Seger R. (2007) The MEK/ERK cascade: from signaling specificity to diverse functions. Biochim Biophys Acta 1773(8):1213-1226.

[267] Liu J, Grimison B, Maller JL. (2007) New insight into metaphase arrest by cytostatic factor: from establishment to release. Oncogene 26(9):1286-1289.

[268] Rauh NR, Schmidt A, Bormann J, Nigg EA, Mayer TU. (2005) Calcium triggers exit from meiosis II by targeting the APC/C inhibitor XErp1 for degradation. Nature 437(7061):1048-1052.

[269] Chun J, Chau AS, Maingat FG, Edmonds SD, Ostergaard HL, Shibuya EK. (2005) Phosphorylation of Cdc25C by pp90Rsk contributes to a G2 cell cycle arrest in Xenopus cycling egg extracts. Cell Cycle 4(1):148-154.

[270] Hutchins JR, Dikovskaya D, Clarke PR. (2003) Regulation of Cdc2/cyclin B activation in Xenopus egg extracts via inhibitory phosphorylation of Cdc25C phosphatase by Ca(2+)/calmodulin-dependent protein [corrected] kinase II. Mol Biol Cell 14(10):4003-4014.

[271] Wang R, Jung SY, Wu CF, Qin J, Kobayashi R, Gallick GE, et al. (2010) Direct roles of the signaling kinase RSK2 in Cdc25C activation during Xenopus oocyte maturation. Proc Natl Acad Sci U S A 107(46):19885-19890.

[272] Wang R, He G, Nelman-Gonzalez M, Ashorn CL, Gallick GE, Stukenberg PT, et al. (2007) Regulation of Cdc25C by ERK-MAP kinases during the G2/M transition. Cell 128(6):1119-1132.

[273] Astuti P, Pike T, Widberg C, Payne E, Harding A, Hancock J, et al. (2009) MAPK pathway activation delays G2/M progression by destabilizing Cdc25B. J Biol Chem 284(49):33781-33788.

[274] Patel R, Holt M, Philipova R, Moss S, Schulman H, Hidaka H, et al. (1999) Calcium/calmodulin-dependent phosphorylation and activation of human Cdc25-C at the G2/M phase transition in HeLa cells. J Biol Chem 274(12):7958-7968.

[275] Planas-Silva MD, Means AR. (1992) Expression of a constitutive form of calcium/calmodulin dependent protein kinase II leads to arrest of the cell cycle in G2. EMBO J 11(2):507-517.

[276] Peng CY, Graves PR, Ogg S, Thoma RS, Byrnes MJ,3rd, Wu Z, et al. (1998) C-TAK1 protein kinase phosphorylates human Cdc25C on serine 216 and promotes 14-3-3 protein binding. Cell Growth Differ 9(3):197-208.

[277] Izumi T, Maller JL. (1995) Phosphorylation and activation of the Xenopus Cdc25 phosphatase in the absence of Cdc2 and Cdk2 kinase activity. Mol Biol Cell 6(2):215-226.

[278] Eymin B, Claverie P, Salon C, Brambilla C, Brambilla E, Gazzeri S. (2006) p14ARF triggers G2 arrest through ERK-mediated Cdc25C phosphorylation, ubiquitination and proteasomal degradation. Cell Cycle 5(7):759-765.

[279] Kotlyarov A, Yannoni Y, Fritz S, Laass K, Telliez JB, Pitman D, et al. (2002) Distinct cellular functions of MK2. Mol Cell Biol 22(13):4827-4835.

[280] Uchida S, Watanabe N, Kudo Y, Yoshioka K, Matsunaga T, Ishizaka Y, et al. (2011) SCFbeta(TrCP) mediates stress-activated MAPK-induced Cdc25B degradation. J Cell Sci 124(Pt 16):2816-2825.

[281] Gutierrez GJ, Tsuji T, Cross JV, Davis RJ, Templeton DJ, Jiang W, et al. (2010) JNK-mediated phosphorylation of Cdc25C regulates cell cycle entry and G(2)/M DNA damage checkpoint. J Biol Chem 285(19):14217-14228.

[282] Goss VL, Cross JV, Ma K, Qian Y, Mola PW, Templeton DJ. (2003) SAPK/JNK regulates cdc2/cyclin B kinase through phosphorylation and inhibition of cdc25c. Cell Signal 15(7):709-718.

[283] Lopez-Aviles S, Grande M, Gonzalez M, Helgesen AL, Alemany V, Sanchez-Piris M, et al. (2005) Inactivation of the Cdc25 phosphatase by the stress-activated Srk1 kinase in fission yeast. Mol Cell 17(1):49-59.

[284] Smith DA, Morgan BA, Quinn J. (2010) Stress signalling to fungal stress-activated protein kinase pathways. FEMS Microbiol Lett 306(1):1-8.

[285] Stanford JS, Lieberman SL, Wong VL, Ruderman JV. (2003) Regulation of the G2/M transition in oocytes of xenopus tropicalis. Dev Biol 260(2):438-448.

[286] Duckworth BC, Weaver JS, Ruderman JV. (2002) G2 arrest in Xenopus oocytes depends on phosphorylation of cdc25 by protein kinase A. Proc Natl Acad Sci U S A 99(26):16794-16799.

[287] Pirino G, Wescott MP, Donovan PJ. (2009) Protein kinase A regulates resumption of meiosis by phosphorylation of Cdc25B in mammalian oocytes. Cell Cycle 8(4):665-670.

[288] Bachmann M, Moroy T. (2005) The serine/threonine kinase Pim-1. Int J Biochem Cell Biol 37(4):726-730.

[289] Mochizuki T, Kitanaka C, Noguchi K, Muramatsu T, Asai A, Kuchino Y. (1999) Physical and functional interactions between Pim-1 kinase and Cdc25A phosphatase. Implications for the Pim-1-mediated activation of the c-Myc signaling pathway. J Biol Chem 274(26):18659-18666.

[290] Wang Z, Bhattacharya N, Mixter PF, Wei W, Sedivy J, Magnuson NS. (2002) Phosphorylation of the cell cycle inhibitor p21Cip1/WAF1 by Pim-1 kinase. Biochim Biophys Acta 1593(1):45-55.

[291] Bachmann M, Hennemann H, Xing PX, Hoffmann I, Moroy T. (2004) The oncogenic serine/threonine kinase Pim-1 phosphorylates and inhibits the activity of Cdc25C-associated kinase 1 (C-TAK1): a novel role for Pim-1 at the G2/M cell cycle checkpoint. J Biol Chem 279(46):48319-48328.

[292] Davezac N, Baldin V, Blot J, Ducommun B, Tassan JP. (2002) Human pEg3 kinase associates with and phosphorylates CDC25B phosphatase: a potential role for pEg3 in cell cycle regulation. Oncogene 21(50):7630-7641.

[293] Mirey G, Chartrain I, Froment C, Quaranta M, Bouche JP, Monsarrat B, et al. (2005) CDC25B phosphorylated by pEg3 localizes to the centrosome and the spindle poles at mitosis. Cell Cycle 4(6):806-811.

[294] Nance J. (2005) PAR proteins and the establishment of cell polarity during C. elegans development.Bioessays 27:126-135.

[295] Rivers DM, Moreno S, Abraham M, Ahringer J. (2008) PAR proteins direct asymmetry of the cell cycle regulators Polo-like kinase and Cdc25. J Cell Biol 180(5):877-885.

[296] Kristjansdottir K, Rudolph J. (2004) Cdc25 phosphatases and cancer. Chem Biol 11(8):1043-1051.

[297] Loffler H, Syljuasen RG, Bartkova J, Worm J, Lukas J, Bartek J. (2003) Distinct modes of deregulation of the proto-oncogenic Cdc25A phosphatase in human breast cancer cell lines. Oncogene 22(50):8063-8071.

[298] Galaktionov K, Chen X, Beach D. (1996) Cdc25 cell-cycle phosphatase as a target of c-myc. Nature 382(6591):511-517.

[299] Dong J, Zeng BH, Xu LH, Wang JY, Li MZ, Zeng MS, et al. (2010) Anti-CDC25B autoantibody predicts poor prognosis in patients with advanced esophageal squamous cell carcinoma. J Transl Med 8:81.

[300] Bugler B, Schmitt E, Aressy B, Ducommun B. (2010) Unscheduled expression of CDC25B in S-phase leads to replicative stress and DNA damage. Mol Cancer 9:29.

[301] Planelles V, Benichou S. (2009) Vpr and its interactions with cellular proteins. Curr Top Microbiol Immunol 339:177-200.

[302] Huard S, Elder RT, Liang D, Li G, Zhao RY. (2008) Human immunodeficiency virus type 1 Vpr induces cell cycle G2 arrest through Srk1/MK2-mediated phosphorylation of Cdc25. J Virol 82(6):2904-2917.

[303] Matsuda N, Tanaka H, Yamazaki S, Suzuki J, Tanaka K, Yamada T, et al. (2006) HIV-1 Vpr induces G2 cell cycle arrest in fission yeast associated with Rad24/14-3-3-dependent, Chk1/Cds1-independent Wee1 upregulation. Microbes Infect 8(12-13):2736-2744.

[304] Elder RT, Yu M, Chen M, Zhu X, Yanagida M, Zhao Y. (2001) HIV-1 Vpr induces cell cycle G2 arrest in fission yeast (Schizosaccharomyces pombe) through a pathway involving regulatory and catalytic subunits of PP2A and acting on both Wee1 and Cdc25. Virology 287(2):359-370.

[305] Frazer C, Young, PG (2012) Carboxy-terminal phosphorylation sites in Cdc25 contribute to enforcement of the DNA damage and replication checkpoints in fission yeast. Curr Genetics , DOI 10.1007/s00294-012-0379-1

Protein Phosphorylation is an Important Tool to Change the Fate of Key Players in the Control of Cell Cycle Progression in *Saccharomyces cerevisiae*

Roberta Fraschini, Erica Raspelli and Corinne Cassani

Additional information is available at the end of the chapter

1. Introduction

Protein phosphorylation is a reversible posttranslational modification that can modulate protein role in several physiological processes in almost every possible way. These include modification of its intrinsic biological activity, subcellular location, half-life and binding with other proteins. Protein phosphorylation is particularly important for the regulation of key proteins involved in the control of cell cycle progression.

Protein phosphorylation is the covalent binding of a phosphate group to some critical residues of the polypeptide. The phosphorylation state of a protein is given by a balance between the activity of protein kinases and protein phosphatases. Eukaryotic protein kinases transfer phosphate groups (PO_4^{3-}) from ATP to an hydroxyl group of the lateral chain of specific serine, threonine or tyrosine residues on peptide substrates. In simple eukaryotic cells, like yeasts, Ser/Thr kinases are more common, while more complex eukaryotic cells, like human cells, have many Tyr kinases. Protein kinases recognize their substrates specifically and their active site consists of an activation loop and a catalytic loop between which substrates bind. Protein kinases differ from each other in the structure of their catalitic domain. A second class of enzymes, protein phosphatases, is responsible for the reverse reaction in which phosphate groups are removed from a protein. Phosphatases gain specificity by binding protein cofactors which facilitate binding to specific phosphoproteins. The active phosphatase often consists of a complex of the phosphatase catalytic subunit and a regulatory subunit.

The use of protein phosphorylation/dephosphorylation as a control mechanism has many advantages since it is rapid, it does not require new proteins to be made or degraded and it is easily reversible. The phosphorylation of specific residues induces structural changes that regulate protein functions by modulating protein folding, substrate affinity, stability and activity. For example, phosphorylation can cause switch-like changes in protein function, which can also lead to major modifications in the catalytic function of enzymes, including kinases and phosphatases. In addition, protein phosphorylation often leads to rearrangement in the structure of the protein that can induce changes in interacting partners or subcellular localization.

Phosphorylation acts as a molecular switch for many regulatory events in signalling pathways that drive cell division, proliferation, differentiation and apoptosis. In order to ensure an appropriate balance of protein phosphorylation, the cell can compartmentalize both protein kinases and phosphatases. Another kind of regulation can be achieved by the spatial distribution of kinases and phosphatases, that creates a gradient of phosphorylated substrates across different subcellular compartments. This spatial separation can also control the activity of other proteins or enzymes and the occurrence of other posttranslational modifications.

Even in a simple organism like budding yeast, approximately 3% of its proteins are kinases or phosphatases. Some of these enzymes are extremely specific, indeed that are able to phosphorylate or dephosphorylate only a few target proteins, while others can act broadly on many proteins. The examples of known targets of phosphorylation include most protein components of the cell like enzymes, structural proteins, cell receptors, ion channels and signaling factors. If a protein is controlled by its phosphorylation state, its activity will be directly dependent on the activity of the kinases and phosphatases that act on it. It is quite common for a phosphate group to be added or removed from a protein continually, a cycle that allows a protein to switch rapidly from one state to another.

A protein can be modified by the addition of a single phosphate group or it can undergo multisite phosphorylation, these events can be driven by a single kinase or by multiple kinases that act in concert. Multisite phosphorylation can determine the extent and duration of a cellular response and can integrate multiple signals on the same protein. For example, many protein kinases involved in the cell cycle control function by generating phosphoSer/Thr-containing sequence motifs in their substrates that are then recognized by phosphoSer/Thr binding proteins. In several cases the phosphopeptide binding domain targets the kinase to prephosphorylated (primed) sites and then mediates processive phosphorylation of the substrate. An important example of this regulation is given by some budding yeast proteins that are phosphorylated by the catalitic subunit of the Cyclin-Dependent Kinase (CDK), Cdc28, that "primes" the protein in order to bind and to be phosphorylated by the Polo-like kinase (Plk) Cdc5.

Cdks are the master regulators of the cell cycle and Cdc5 plays key roles during all stages of mitosis and in the cell cycle checkpoint response to genotoxic stress [1,2]. The protein kinase Cdc28 drives every cell cycle transition and its activity is tightly regulated by phosphorylation/dephosphorylation events and by its association with a class of proteins called cyclins. Cyclins levels fluctuate during the cell cycle and their binding to Cdc28 activates its catalitic domain and allows its binding to specific substrates. Polo-like kinases have a conserved Ser/Thr kinase domain and a noncatalytic C-terminal region composed by two homologous boxes Polo Box Domain (PBD) [3], that appear to target the kinase to mitotic substrates. In addition, it has been shown that phosphopeptide binding to the PBD stimulates kinase activity, suggesting a conformational switching mechanism for Plk regulation and a double function for the PBD.

It is the simplicity, reversibility and flexibility of phosphorylation that explains why it has been adopted as the most general control mechanism of the cell. Below we describe how phosphorylation and dephosphorylation events can finely regulate in space and time some key proteins in the control of cell cycle progression in *Saccharomyces cerevisiae* .

2. The protein kinase Swe1 is regulated at several levels

Budding yeast cyclin-dependent kinase Cdk1 is the motor that drives cell cycle progression. Cdk1 activity is regulated at multiple levels in different cell cycle phases by interactions with different proteins, called cyclins, and by phosphorylation and dephosphorylation events. In G2 phase of the cell cycle, the protein kinase Swe1 inhibits Cdk1 by phosphorylating the conserved Y19 residue of its catalytic subunit Cdc28 [4], thus blocking both switch from polar to isotropic bud growth and nuclear division, since both these events rely on G2/M Cdk activity. This inhibitory phosphorylation is reversed by the Mih1 phosphatase [5], leading to Cdk1 activation and entry into mitosis. Swe1 phosphorylates and inhibits Cdc28 during every cell cycle and in case of problems in the bud neck, bud formation, actin cytoskeleton and abnormal cell shape. Successful bud formation leads to Swe1 degradation in late G2 phase, while morphological defects block this degradation, thus delaying entry into mitosis. Swe1 has therefore a critical role in coordinating cell morphogenesis with nuclear division.

Swe1 levels are controlled by the "morphogenesis checkpoint", a pathway that is activated in response to alterations in the actin cytoskeleton or in septin organization. The activation of this checkpoint ultimately leads to Swe1 stabilization and a subsequent delay in nuclear division [6]. In an unperturbed cell cycle, Swe1 is recruited to the mother-bud neck in S phase; this change in its localization, which is promoted by the interaction with its regulators Hsl1 and Hsl7, is essential for subsequent Swe1 multiphosphorylation, an event that leads to its ubiquitylation and degradation, thus allowing entry into mitosis. The morphogenesis checkpoint causes Swe1 stabilization by interfering with its localization to the bud neck, acting directly on Swe1 or on its regulators Hsl1 and Hsl7, thus preventing modifications that lead to its degradation [7]. Accordingly, the lack of septin localization at the bud neck results in Swe1 stabilization [7-9], and even subtle perturbations in septin structure interfere with Hsl1 and Swe1 localization to the bud neck [7]. Moreover, actin depolymerization in budded cells causes both stabilization of Swe1 and its displacement from the bud neck without altering Hsl1 localization [7], indicating that morphogenesis checkpoint activation prevents Swe1 degradation by interfering with Swe1 localization to the bud neck, thus inhibiting its phosphorylation.

Swe1 is subjected to multiple regulations that change its phosphorylation state (35-40 phosphorylated sites have been identified *in vivo*), subcellular localization and protein levels (Figure 1). During S phase, Swe1 accumulates in the nucleus where it is phosphorylated by Clb-Cdc28 before being exported to the cytoplams and then to the daughter side of the bud neck. Swe1 nuclear export is essential for its degradation in G2/M in fact a Swe1 variant that cannot be exported from the nucleus is largely stabilized [10]. Clb-Cdc28 phosphorylates Swe1 at multiple sites [11] and these events can occur in the nucleus, in the cytoplasm or at the bud neck. This multiphosphorylation by Cdc28-Clb has different roles: on one hand it promotes Swe1 activity generating a feedback loop, while on the other hand it promotes Swe1 degradation [12]. At the mother-bud neck, filaments of conserved proteins called septins (Cdc3, Cdc10, Cdc11, Cdc12 and Shs1 in *S. cerevisiae*) form a dynamic ring structure [13] that is essential for the recruitment of a number of proteins involved in the control of cell cycle progression [14,15]. Among these, the septin ring acts as a platform to recruit several Swe1 regulators, such as the Hsl1 protein kinase and its adaptor Hsl7, both essential for Swe1 localization to the bud neck and for its phosphorylation [7,9]. Hsl1, whose kinase activity requires assembled septins, undergoes autophosphorylation [8,16] and phosphorylates Hsl7 [9] but does not seem to phosphorylates Swe1 [17] although its kinase activity is required for Swe1 recruitment at the bud neck [18]. The fact that Hsl1 activation requires assembled septins ensures that Swe1 degradation does not begin until a bud has formed, thus providing a link between bud formation and entry into mitosis [18]. Also the PAK (p21-activated kinase) kinase Cla4 associates with the septin ring and is involved in Swe1 phosphorylation, probably during S phase [19]. Moreover Swe1 phosphorylation by Clb-Cdc28 promotes subsequent phosphorylation by the Polo-like kinase Cdc5; in fact the synergistic phosphorylation that can be observed *in vitro* on Swe1 by Clb2-Cdc28 and Cdc5 is the result of priming Swe1 by Cdc28, in which the resulting phosphorylated Swe1 becomes a better substrate for Cdc5 [20]. An additional level of Swe1 regulation involves the Cdc55 regulatory subunit of protein phosphatase PP2A, that has been implicated in its degradation since loss of Cdc55 function causes Swe1 stabilization [21]. However, how this control happens at the molecular level is not clear. In addition, a mathematical model for the morphogenesis checkpoint activation suggested that a subset of Swe1 phosphorylations could inhibit its activity whereas other phosphorylations could target Swe1 for degradation [22]. In any case, hyperphosphorylated Swe1 species are recognized by a still unidentified ubiquitin ligase and ubiquitylated [23]. Subsequently, Swe1 is degraded via the proteasome and this event allows mitotic entry [23]. However, how bud neck-localized Swe1 is targeted to degradation after phosphorylation is still obscure. There are Swe1 variants that do not undergo degradation although they show proper bud neck localization, phosphorylation and interaction with known Swe1 regulators [23], indicating that still unknown Swe1 regulators exist.

So, Swe1 regulation is governed by complex pathways that are still partially unidentified. In particular, the involvement of the ubiquitylation pathway and the identity of the related ubiquitin ligase(s) involved in Swe1 degradation are unknown. A possible role in this control can be hypothesized for the functionally redundant budding yeast proteins Dma1 and Dma2, which belong to the same FHA-RING ubiquitin ligase family as *S. pombe* Dma1 and human Chfr and Rnf8. FHA domains are phosphothreonine-binding modules [24] frequently found in DNA repair and checkpoint proteins [25,26] and RING domains are typical of E3 ubiquitin ligases [26]. The presence of an FHA domain implies that one or more protein kinases function upstream of these proteins. An intact FHA domain is required for checkpoint function of all characterized FHA-RING ubiquitin ligases. All these proteins appear to control different aspects of the mitotic cell cycle, but several molecular details of their functions are still obscure [for review, see 27]. In particular, the *S. cerevisiae* Dma proteins are involved in mitotic checkpoints and contribute to control septin ring dynamics and cytokinesis by unknown mechanisms [28,29]. Moreover, *in vitro* ubiquitin ligase activity of Dma1 and Dma2 has been described [30], although their *in vivo* targets are still unknown.

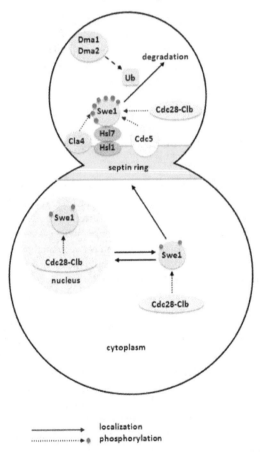

localization
phosphorylation

Figure 1. Model for Swe1 regulation. Swe1 shuttles from the nucleus to the cytoplasm and then it is translocated to the septin ring at the bud neck. These events are controlled through phosphorylation by Cdc28-Clb kinase. Once at the bud neck, Swe1 undergoes other phosphorylation events that drive its ubiquitylation and ultimately lead to its degradation via the proteasome. Dashed line indicates that Dma1 and Dma2 ubiquitin ligases are involved, directly or indirectly, in Swe1 ubiquitylation.

Recently, we provided genetic and biochemical evidence that Dma proteins are involved in promoting Swe1 ubiquitylation *in vivo* and contribute to the regulation of Swe1 stability by acting in a step following the recruitment of Swe1 to the bud neck and its phosphorylation [31]. Indeed, the lack of Dma proteins leads to accumulation of fully phosphorylated and bud neck localized Swe1. However, as *dma1Δ dma2Δ* cells are viable while Swe1 degradation is essential for cell viability, other yet unidentified ubiquitin ligases likely ubiquitylate Swe1 during an unperturbed cell cycle. The Dma-dependent Swe1 down-regulation, whose lack does not significantly affect unperturbed cell cycle progression when the other Swe1 regulatory pathways are proficient, appears to be crucial for proper response to DNA replication stress.

Collectively, the complex Swe1 regulation is an example of how phosphorylation events drive the fate of a protein. Indeed, Swe1 undergoes multiple phosphorylations that are crucial for its localization, its activity and its interaction with an E3 ubiquitin ligase that promote its degradation; moreover also some Swe1 regulators are regulated by phosphorylation, showing how this modification is largely involved in the control of cellular events.

3. The complex regulation of the protein kinase Kin4, a key player of the spindle position checkpoint

The spindle position checkpoint (SPOC) is the important pathway that blocks mitotic exit and cytokinesis in case of mitotic spindle misalignment [32]. This pathway is crucial for budding yeast since the division site is determined early in the yeast cell cycle, in late G1 concomitantly with bud site selection, while the mitotic spindle is assembled after this event. So, in order to allow proper chromosome distribution between mother and daugher cell, the mitotic spindle must be correctly aligned with respect to the mother bud axis before mitotic exit. The SPOC monitors this event and arrests mitotic progression in case of spindle misalignment. The target of the SPOC is the GTPase Tem1, that acts at the top of the Mitotic Exit Network (MEN). The MEN is a signaling cascade of protein kinases that controls both exit from mitosis and cytokinesis. At the top of the pathway, the GTPase Tem1 activates Cdc15 kinase that leads to Dbf2/Dbf20 and Mob1 activation. The MEN ultimately activates the protein phosphatase Cdc14 that leads to Cdk1 inactivation, a key event for both mitotic exit and cytokinesis [33]. The localization of all these proteins is important for their activity. The spindle pole body (SPB) component Nud1 is the platform for most MEN components localization to the spindle poles during mitosis [34]. During every cell cycle, Tem1 is kept inactive until the mitotic spindle is properly aligned respect to the mother-bud axis, thus coupling mitotic exit with nuclear division [35-37]. Tem1 regulation is complex: it is positively regulated by Lte1 [35] and it is kept inactive by the dimeric GTPase-activating protein (GAP) Bub2-Bfa1, which is the target of two protein kinases, Cdc5 and Kin4 [38-41]. Cdc5 is the budding yeast Polo-like kinase, it plays multiple functions in mitosis and cytokinesis through phosphorylation of different substrates. Cdc5 inhibits Bfa1 by phosphorylation at the anaphase onset [40,41], thus promoting timely Tem1 activation and mitotic exit. Kin4 is a non essential serine/threonine protein kinase that plays only a minor role in mitotic progression in normal growth conditions. But, importantly, Kin4 kinase participates in the SPOC and indeed it is essential to delay cell cycle progression of cells with a misaligned spindle. In case of spindle mispositioning, Kin4 maintains Bub2-Bfa1 GAP complex active through phosphorylation of Bfa1 on residues Ser150 and Ser180 [42], these events counteract Cdc5 action on Bfa1 and thereby inhibit mitotic exit.

As already said, the subcellular localization of MEN and SPOC components is critical for their function, indeed the asymmetric distribution of MEN activators and inhibitors is one element that couples mitotic exit with correct nuclear migration. During an unperturbed cell cycle the Bub2-Bfa1 complex and Tem1 preferentially associates with the SPB that enters the daughter cell (dSPB) [35,37]. Instead the Lte1 protein, which positively regulates Tem1, is confined in the bud from the G1/S transition to telophase, when it diffuses throughout the cytoplasm of both mother cell and bud [37]. During anaphase, the SPB-associated Bub2-Bfa1 GAP complex keeps Tem1 inactive until the SPB and spindle enter the bud, where Tem1 is activated by its encounter with Lte1, thus promoting mitotic exit. Similarly to MEN factors, Kin4 binds the SPB in a Nud1 dependent manner [34,43]. During normal cell cycle progression, Kin4 localizes to the cortex of the mother cell by interaction of its C-terminal regulatory domain and this binding is a prerequisit for its loading on mother-bound SPB (mSPB) during midanaphase [44]. When spindles is correctly oriented, Kin4 and Bub2-Bfa1 are asymmetrically localized to opposite SPBs. With anaphase onset, Bub2-Bfa1 then becomes inhibited by the Cdc5 kinase thus promoting mitotic exit. On the contrary, in case of spindle misalignment, Kin4 and Bub2-Bfa1 are brought together at both SPBs [45-48]; in these conditions Kin4 prevents Bfa1 phosphorylation by Cdc5, thereby inhibiting mitotic exit and this regulation is essential for survival of cells with a misaligned spindle.

So, Kin4 subcellular localization and activity are finely regulated during an unperturbed cell cycle and in response to spindle misalignment and these events involve its phosphorylation state (Figure 2). Kin4 phosphorylation state changes during the cell cycle: it is hyperphosphorylated during late stages of mitosis but is hypophosphorylated during S-phase and early mitosis [49]. At present, the relationship between Kin4 function, phosphorylation and localization is not fully understood, but at least three regulators of Kin4 have been identified: Lte1, the protein phosphatase PP2A-Rts1 and the bud neck kinase Elm1.

Recent data indicate that Lte1 can regulate Kin4 by controlling its phosphorylation status [50]: Lte1 binds Kin4 and promotes its hyperphosphorylation and this event restricts Kin4 binding to the mSPB and prevents Kin4 that escapes the mother cell from associating with the dSPB. Importantly, this Lte1-mediated exclusion of Kin4 from the dSPB is essential for proper mitotic exit of cells with a correctly aligned spindle. Therefore, Lte1 promotes mitotic exit by inhibiting Kin4 activity at the dSPB. However, how this regulation happens remains to be determined. Indeed, Lte1 might regulate Kin4 activity itself, the Kin4 phosphatase PP2A-Rts1, the Kin4 kinase Elm1, other Kin4 kinases or the availability of Kin4 itself to be phosphorylated. So the SPOC function could be linked to the spatial restriction of the MEN regulators Kin4 and Lte1 and inhibition of Kin4 by Lte1 [50].

On the basis of the correlation between phosphorylation status and time of Kin4 activation during the cell cycle, it was hypothesized that dephosphorylated Kin4 might be its active form and that phosphatases that promote accumulation of this form would be required for Kin4 function. Recently, the protein phosphatase PP2A and its regulatory subunit Rts1 have been identified as Kin4 regulators. In particular, the phosphatase PP2A-Rts1 is required for Kin4 dephosphorylation during cell cycle entry and to maintain Kin4 in the dephosphorylated state during S phase and mitosis [49]. In addition, PP2A-Rts1 is crucial for proper localization of Kin4 to the mother cortex and SPBs both during the cell cycle and in response to spindle position defects [49]. The phosphatase does not appear to affect Kin4 kinase activity but instead it promotes its association with SPBs, which is essential for SPOC activity. So, PP2A-Rts1 functions upstream of Kin4, regulating its phosphorylation and localization during an unperturbed cell cycle and during SPOC activation, thus defining the phosphatase as a key regulator of SPOC function [49]. However, it is still unclear how Rts1 mobilizes Kin4 at the cortex and enables its association with SPBs, indeed it might dephosphorylate Kin4 or a Kin4 interactor at SPBs.

Another key player of Kin4 regulation is the bud neck-localized Elm1 kinase [51,52]. Elm1 has been previously implicated in septin organization, bud morphogenesis, bud neck integrity and cell cycle progression [53-55]. Recently it has been shown that Elm1 contributes to the SPOC by promoting Kin4 kinase activity. Indeed, Elm1 activates Kin4 by direct phosphorylation on Thr209, a residue in the activation loop, and on additional sites at the C terminus of Kin4 [51,52]. Kin4 phosphorylation by Elm1 is critical for its kinase activity and subsequent hyperphosphorylation [49,51,52]. At present, the molecular mechanism by which phosphorylation of Thr209 influences Kin4 catalytic function is unclear. Structural studies of kinases regulated by activation loop phosphorylation indicate that the loop might function as a "sensitive switch" that controls substrate binding triggered by phosphorylation-dependent conformational changes [56]. However, the lack of Kin4 kinase activity and hyperphosphorylation observed in elm1Δ cells lead to reduced Kin4 binding to the mother cell cortex and SPOC deficiency. So, Elm1 function in the SPOC upstream of Kin4 by controlling its activity and localization.

Interestingly, we have recently described a new level of complexity in SPOC regulation. Indeed, we have implicated the E3 ubiquitin ligases Dma1 and Dma2 in the control of Elm1 bud neck localization [29]. Elm1 mislocalization in dma1Δ dma2Δ cells results in reduced levels of Kin4 kinase activity and asymmetry of Bub2-Bfa1 at the spindle poles of mispositioned spindles. So, Dma1, Dma2, Elm1 and Kin4 are part of the same SPOC regulatory module, with Dma proteins controlling Elm1 localization, that is in turn required for full Kin4 activation. Importantly, it is worth noting that Dma1, Dma2, Elm1 and Kin4 all are required to prevent mitotic exit in response to spindle mispositioning but not to spindle depolymerization, differently from Bub2 and Bfa1 [28,39,52].

NORMAL ANAPHASE   **MISPOSITIONED SPINDLE**

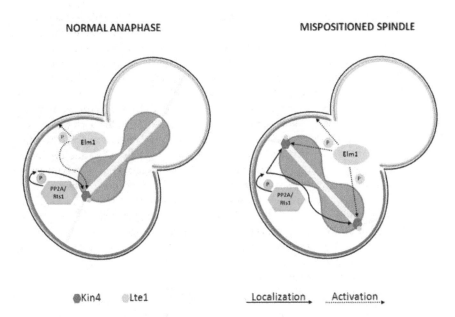

●Kin4   ◦Lte1   Localization   Activation

Figure 2. Model for Kin4 regulation. Kin4 is localized to the mother cell cortex and to the mother bound spindle pole body (SPB) when the mitotic spindle is correctly oriented. On the contrary, in case of spindle misalignment, Kin4 is present at both SPBs. The protein phosphatase PP2A-Rts1 is crucial for the proper localization of Kin4 to the mother cortex and SPBs both during the cell cycle and in response to spindle position defects. The protein kinase Elm1 activates Kin4 by direct phosphorylation and controls its localization.

4. Budding yeast cytokinesis is controlled by several phosphoproteins

Cytokinesis is the spatially and temporally regulated process by which, after chromosome segregation, eukaryotic cells divide their cytoplasm and membranes in order to produce two daughter cells. In budding yeast, the activity of the components of the citokinetic machinery is tightly controlled and their recruitment to the mother-bud neck follows a hieralchical order of assembly during the cell cycle. Most of these regulations are driven by phosphorylation and dephosphorylation events.

Cytokinesis completion is driven by complex and partially redundant pathways that regulate the assembly and contraction of an actomyosin ring (AMR) and the deposition of a trilaminary septum between mother and bud. In particular, the AMR is involved in constricting the plasma membrane at the division site to complete closure [57,58] and AMR contraction is coupled to the centripetal growth of the primary septum (PS).

The first step towards cytokinesis is the assembly of a septin ring, which forms at the bud neck concomitantly with bud emergence as soon as cells enter S phase, and marks the position where constriction between mother and daughter cell will take place at the end of mitosis. During S phase, the septin ring becomes an hourglass shaped structure that serves as a scaffold for recruiting other proteins to the bud neck, among which the type II myosin heavy chain Myo1 that forms a ring at the presumptive bud site during early S phase [57]. This Myo1 ring persists at the mother-bud neck until the end of anaphase, when a coincident ring of F-actin assembles and the resulting AMR eventually contracts, accomplishing septum formation. In late anaphase Hof1, Cyk3 and Inn1 are recruited sequentially to the bud neck. All these proteins are required to activate the chitin synthase Chs2 [59-61], which is in turn recruited to the division site after mitotic exit and is required to build the primary septum (PS) which is mostly made of chitin. Once the cytokinetic apparatus is fully assembled, AMR contraction, membrane invagination and PS synthesis all begin almost immediately. After completion of the PS, at either side of this structure, secondary septa (SS), which are made of the same components as the cell wall, i. e. glucans and mannan, are synthetized [62]. At this point, mother and daughter cells are connected by a trilaminar septum. Afterwards, cell separation is driven by the action of endochitinase, Cts1, and glucanases, Dse2 Dse4 Egt2, that degrade the PS from the daughter side [63,64]. Then mother and daughter cells separate permanently from each other leaving a disk of chitin, called "bud scar", on the mother cell surface.

The events leading to cytokinesis must be tightly controlled and coordinated with chromosome segregation and mitotic exit in order to ensure the genetic stability during cell growth and thereby the fate of daughter cells. Several pathways are able to regulate the last event of the cell cycle including the Mitotic Exit Network (MEN). The MEN seems to control not only the exit from mitosis but also the timing of cytokinesis. In fact, several MEN components localize to the division site after mitotic exit and they likely play a direct role in the regulation of cytokinesis. Indeed, as mitotic Cdk1 activity decrease, MEN kinases Cdc15, Dbf2/Dbf20, Mob1, Cdc5 and the protein phosphatase Cdc14 associate with the bud neck [65-69], in addition some MEN mutants fail to undergo cytokinesis [70,71]. Here we describe examples of regulation by phosphorylation and/or dephosphorylation during cytokinesis catalyzed by Cdc28-Clb2 and/or Dbf2-Mob1 kinases and also by Cdc14 phosphatase, and how these post-translational modifications can regulate the function of three important cytokinetic proteins, Cyk3, Hof1 and Chs2, in space and time.

Cyk3 is an SH3-domain protein and was isolated as high-copy suppressor of lethality in an *iqg1Δ* strain [72] and in a *myo1Δ* strain [73]. Cyk3 interacts with Hof1 and both cooperate to recruit Inn1 [61,74], that is essential for activation of chitin synthase and therefore for primary septum formation [60,61]. Overexpression of *CYK3* leads to an actomyosin independent recruitment of Inn1 to the bud neck [74], indicating that Cyk3 plays a central role in a rescue mechanism for cytokinesis in the absence of a functional AMR. Cyk3 activity and localization are positively regulated by phosphorylation events (Figure 3). Cyk3 total levels are constant throughout the cell cycle, but interestingly Cyk3 phosphorylated species begin to accumulate after mitotic exit and this leads to its recruitment to the bud neck. This phosphorylation event requires the MEN activity [75]. In particular, it has been proposed that Cyk3 is phosphorylated by Dbf2-Mob1 kinase but if it is its direct substrate has not been determined.

It has instead been demonstrated that the Dbf2-Mob1 kinase, that appears at the division site just before AMR contraction [76,77], directly phosphorylates Hof1, a member of the F-BAR (Fes/CIP4 homology Bin/Amphiphysin/Rvsp) protein family conserved from yeast to mammals [78]. Hof1 protein levels, activity and localization are regulated by many phosphorylation events. Hof1 protein levels are regulated during the cell cycle, in particular Hof1 accumulates from G1/S phase, it disappears after mitotic exit, concomitantly with AMR contraction, and it remains unstable during the G1 phase of the following cell cycle. Hof1 colocalizes with the septin ring from S phase to late anaphase, then it interacts with the AMR before its contraction [79]. Efficient AMR contraction and cell separation are allowed by subsequent degradation of Hof1 by the action of the E3 ubiquitin ligase complex SCF (Skp, Cullin, F-box containing complex)/Grr1 [80] that can recognize its PEST (Proline, glutamin acid (E), Serine and Threonine) domain. Hof1 is directly phosphorylated by three kinases that act in concert to regulate its activity: Clb2-Cdc28, Cdc5 and Dbf2-Mob1 [81]. During mitosis, Hof1 undergoes Clb2-Cdc28 phosphorylation that does not seem to control its localization, but it primes Hof1 for subsequent phosphorylation by Cdc5 at Ser517. This event facilitates Hof1 binding to Mob1. The Dbf2-Mob1 kinase then phosphorylates several residues at both N- and C- terminal of Hof1 [81]. These modifications do not influence the physical interaction between Hof1 and Cyk3 or Inn1, instead they promote Hof1 release from the septin ring and localization to the AMR (Figure 3). There, phosphorylated Hof1 promotes AMR contraction by an unknown mechanism.

The third example is the regulation of the chitin synthase that builds the PS. Chs2 is an integral membrane protein that polymerizes chitin from its precursor UDP-N-acetyl-glucosamine. The chitin synthase 2 is synthetized in G2/M and accumulates in the endoplasmic reticulum (ER). The N-terminal and the central catalytic domain are located in the cytoplasm while the hydrophobic C-terminal domain is integrated into the ER membrane. During the metaphase to anaphase transition, Chs2 is phosphorylated in several Ser-Pro sites at its N-terminus exposed into the cytoplasm by Cdc28-Clb2 [82,83]. These phosphorylations events are important for Chs2 retention into the ER and therefore for its inibition. At the end of mitosis, MEN activation leads to the decrease of Cdc28-Clb2 activity and Chs2 can be exported from the ER and targeted to the division site [75,84-86]. The molecular mechanism of this release was unclear, but recent data indicate that Chs2 N-terminus is directly dephosphorylated by the protein phosphatase Cdc14 [86]. Then dephosphorylated Chs2 is traslocated to Golgi and, through secretory vescicoles, is delivered to the plasma membrane at the bud neck (Figure 3) [85,87]. There, Chs2 is active and promotes centripetally deposition of PS that occurs concomitantly with AMR contraction and membrane ingression. After PS completion Chs2 is removed from the bud neck by endocytosis and transferred to the vacuole where it is degraded by the action of the protease Pep4 [87]. The signal that leads to Chs2 endocytosis is currently unknown.

In summary, Cyk3, Hof1 and Chs2 regulations are good examples of how phosphorylation events can change the fate of a protein and underline the relevance of this posttranslational modification in the control of a very important cell cycle process, the cytokinesis.

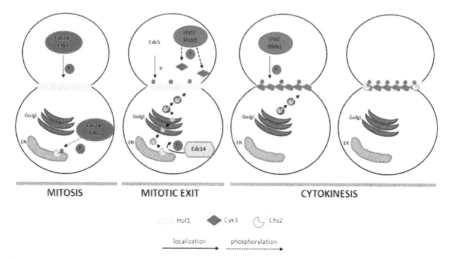

Figure 3. Model for Cyk3, Hof1 and Chs2 regulation. Cyk3 is likely phosphorylated by Dbf2-Mob1 kinase (dashed line) and recruited to the bud neck. Septin bound Hof1 is "primed" by Cdc28-Clb2 phosphorylation and then it is phosphorylated by polo-like kinase Cdc5 and by Dbf2-Mob1 kinase. These modifications promote Hof1 release from the septin ring and localization to the actomyosin ring. Chs2 accumulates in the endoplasmic reticulum (ER) with its N-terminus exposed into the cytoplasm, this tail is phosphorylated by Cdc28-Clb2. These phosphorylations events are important for Chs2 retention into the ER. At the end of mitosis, Chs2 N-terminus is directly dephosphorylated by the protein phosphatase Cdc14, then dephosphorylated Chs2 is traslocated to Golgi and, through secretory vescicoles, is delivered to the plasma membrane at the bud neck.

5. Conclusion

The phosphorylation of a protein is a simple mechanism that alters its conformation, and so its ability to function, in a reversibile way. As we can learn from the examples of protein regulation that we have focused on, phosphorylation is a flexible mechanism that regulates the target protein in several ways. Indeed, phosphorylation is not simply used to switch the activity of a protein on or off, but can have many additional roles. It can influence its ability to form complexes with other proteins, it can affect the rate at which a protein is degraded or its ability to localize to a particular subcellular location. For example, phosphorylation events on the protein kinases Swe1 and Kin4, and on the cytokinetic proteins Cyk3, Hof1 and Chs2 lead to change in their localization.

The action of kinases is counteracted by phosphatases and both controls are essential to determine the phosphorylation state of the target proteins. The balance of phosphorylation and dephosphorylation can also be critical in determining the strength and duration of the response. Therefore, kinases and phosphatases must be regulated spatiotemporally in order to obtain the proper cellular response.

In addition, a protein can be modified by the addition of a single phosphate group or by multiple phosphates, by a single protein kinase or by multiple kinases. Multisite phosphorylation is a strategy that enables two or more effects to operate in the same protein. Indeed, some phosphorylation events "prime" the protein in order to be phosphorylated by another kinase that acts subsequently in the same cellular compartment or in another location. The protein kinase Swe1 is a very good example of this kind of regulation, in fact several kinases act sequentially leading to the accumulation of hyperphosphorylated bud neck localized Swe1. Also the cytokinetic proteins Hof1 and Chs2 are the target of at least two different protein kinases that change Hof1 and Chs2 activity and localization. Alternatively, a phosphorylation event can inhibit the phosphorylation of other residues on the same protein. For example, Bfa1 phosphorylation by Kin4 inhibits phosphate group addition by the polo-like kinase Cdc5 and so keeps Bfa1 active.

Another important issue is the crosstalk between phosphorylation and other posttranslational modifications. Crosstalks can be positive or negative, thus promoting or inhibiting the subsequent modification. About phosphorylation and ubiquitylation, an increasing number of phosphoproteins are then ubiquitylated. Both Hof1 and Swe1 are hyperphosphorylated and afterwards ubiquitylated, subsequently they are targeted to degradation via the proteasome. A very interesting issue is how the specificity is determined. Indeed, phosphorylated residues of serine, threonine and tyrosine recognition by phosphobinding domains depends on the sequence of amino acids immediately around the phosphorylated residue, whereas recognition of monoUb by ubiquitin binding domains does not seem to be influenced by the primary sequence bearing the ubiquitinated lysine.

Even if there are several open questions regarding the molecular mechanisms that control the phosphoproteins that we have described, they are good examples of how the phosphorylation event can change the fate of a protein.

Author details

Roberta Fraschini*, Erica Raspelli[†] and Corinne Cassani
Università degli Studi di Milano-Bicocca, Dipartimento di Biotecnologie e Bioscienze, Milano, Italy

Acknowledgement

Roberta Fraschini's research is supported by grants from PRIN (Progetti di Ricerca di Interesse Nazionale) 2008. C. C. is supported by a fellowship from Fondazione Confalonieri.

6. References

[1] Morgan DO (2007) The Cell Cycle: Principles of Control. Oxford University Press.

---

* Corresponding Author
† Equally Contributing Authors

[2] Nigg EA (1998) Polo-like kinases: positive regulators of cell division from start to finish. Curr Opin Cell Biol. 10:776-83.

[3] Sonnhammer EL, Eddy SR, Birney E, Bateman A, Durbin R (1998) Pfam: multiple sequence alignments and HMM-profiles of protein domains. Nucleic Acids Res. 26: 320–322.

[4] Booher RN, Deshaies RJ, Kirschner MW (1993) Properties of *Saccharomyces cerevisiae* wee1 and its differential regulation of p34CDC28 in response to G1 and G2 cyclins. EMBO J. 12: 3417-26.

[5] Russell P, Moreno S, Reed SI (1989) Conservation of mitotic controls in fission and budding yeasts. Cell. 57: 295-303.

[6] Lew DJ (2003) The morphogenesis checkpoint: how yeast cells watch their figures. Curr Opin Cell Biol 15: 648-53.

[7] Longtine MS, Theesfeld CL, McMillan JN, Weaver E, Pringle JR, Lew DJ (2000) Septin-dependent assembly of a cell cycle-regulatory module in *Saccharomyces cerevisiae*. Mol Cell Biol. 20: 4049-61.

[8] Barral Y, Parra M, Bidlingmaier S, Snyder M (1999) Nim1-related kinases coordinate cell cycle progression with the organization of the peripheral cytoskeleton in yeast. Genes Dev 13: 176-87.

[9] Shulewitz MJ, Inouye CJ, Thorner J (1999) Hsl7 localizes to a septin ring and serves as an adapter in a regulatory pathway that relieves tyrosine phosphorylation of Cdc28 protein kinase in *Saccharomyces cerevisiae*. Mol Cell Biol.19: 7123-37.

[10] Keaton MA, Szkotnicki L, Marquitz AR, Harrison J, Zyla TR, Lew DJ (2008) Nucleocytoplasmic trafficking of G2/M regulators in yeast. Mol Biol Cell. 19: 4006-18.

[11] Harvey SL, Charlet A, Haas W, Gygi SP, Kellogg DR (2005) Cdk1-dependent regulation of the mitotic inhibitor Wee1. Cell. 122: 407-20.

[12] Keaton MA, Lew DJ (2006) Eavesdropping on the cytoskeleton: progress and controversy in the yeast morphogenesis checkpoint. Curr Opin Microbiol. 9: 540-6.

[13] Versele M, Thorner J (2005) Some assembly required: yeast septins provide the instruction manual. Trends Cell Biol. 15: 414-24.

[14] Longtine MS, Bi E (2003) Regulation of septin organization and function in yeast. Trends Cell Biol. 13: 403-9.

[15] Gladfelter AS, Kozubowski L, Zyla TR, Lew DJ (2005) Interplay between septin organization, cell cycle and cell shape in yeast. J Cell Sci. 118: 1617-28.

[16] McMillan JN, Longtine MS, Sia RA, Theesfeld CL, Bardes ES, Pringle JR, Lew DJ (1999) The morphogenesis checkpoint in *Saccharomyces cerevisiae*: cell cycle control of Swe1p degradation by Hsl1p and Hsl7p. Mol Cell Biol. 19: 6929-39.

[17] Cid VJ, Shulewitz MJ, McDonald KL, Thorner J (2001) Dynamic localization of the Swe1 regulator Hsl7 during the *Saccharomyces cerevisiae* cell cycle. Mol Biol Cell.12: 1645-69.

[18] Theesfeld CL, Zyla TR, Bardes EG, Lew DJ (2003) A monitor for bud emergence in the yeast morphogenesis checkpoint. Mol Biol Cell. 14: 3280-91.

[19] Sakchaisri K, Asano S, Yu LR, Shulewitz MJ, Park CJ, Park JE, Cho YW, Veenstra TD, Thorner J, Lee KS (2004) Coupling morphogenesis to mitotic entry. Proc Natl Acad Sci U S A. 101: 4124-9.

[20] Asano S, Park JE, Sakchaisri K, Yu LR, Song S, Supavilai P, Veenstra TD Lee KS (2005). Concerted mechanism of Swe1/Wee1 regulation by multiple kinases in budding yeast. EMBO J. 24: 2194–2204.

[21] Yang H, Jiang W, Gentry M, Hallberg RL (2000) Loss of a protein phosphatase 2A regulatory subunit (Cdc55p) elicits improper regulation of Swe1p degradation. Mol Cell Biol. 20: 8143-56.

[22] Ciliberto A, Novak B, Tyson JJ (2003) Mathematical model of the morphogenesis checkpoint in budding yeast. J Cell Biol.163: 1243-54.

[23] McMillan JN, Theesfeld CL, Harrison JC, Bardes ES, Lew DJ (2002). Determinants of Swe1p degradation in Saccharomyces cerevisiae. Mol Biol Cell.13: 3560-75.

[24] Hofmann K, Bucher P (2005) The FHA domain: a putative nuclear signalling domain found in protein kinases and transcription factors. Trends Biochem Sci. 20:347-9.

[25] Durocher D, Jackson SP (2002) The FHA domain. FEBS Lett 513: 58–66.

[26] Lorick KL, Jensen JP, Fang S, Ong AM, Hatekeyama S, Wissman AM (1999) RING fingers mediate ubiquitin-conjugating enzyme (E2)-dependent ubiquitination. PNAS 28: 11364-9.

[27] Brooks L 3rd, Heimsath EG Jr, Loring GL, Brenner C (2008) FHA-RING ubiquitin ligases in cell division cycle control. Cell Mol Life Sci. 65: 3458-66.

[28] Fraschini R, Bilotta D, Lucchini G, Piatti S (2004) Functional characterization of Dma1 and Dma2, the budding yeast homologues of Schizosaccharomyces pombe Dma1 and human Chfr. Mol Biol Cell 15: 3796-810.

[29] Merlini L, Fraschini R, Boettcher B, Barral Y, Lucchini G, Piatti S (2012) Budding yeast Dma proteins control septin dynamics and the spindle position checkpoint by promoting the recruitment of the Elm1 kinase to the bud neck. Plos Genetics, in press.

[30] Loring GL, Christensen KC, Gerber SA, Brenner C (2008) Yeast Chfr homologs retard cell cycle at G1 and G2/M via Ubc4 and Ubc13/Mms2-dependent ubiquitination. Cell Cycle 7: 96-105.

[31] Raspelli E, Cassani C, Lucchini G, Fraschini R (2011) Budding yeast Dma1 and Dma2 participate in regulation of Swe1 levels and localization. Mol. Biol. Cell. 22: 2185-2197.

[32] Fraschini R, Venturetti M, Chiroli E, Piatti S (2008) The spindle position checkpoint: how to deal with spindle misalignment during asymmetric cell division in budding yeast. Biochemical Society Transactions Volume 36, part 3, 416-420.

[33] Stegmeier F, Amon A (2004) Closing mitosis: the functions of the Cdc14 phosphatase and its regulation. Annu Rev Genet. 38: 203-32.

[34] Gruneberg U, Campbell K, Simpson C, Grindlay J, Schiebel E (2000) Nud1p links astral microtubule organization and the control of exit from mitosis. EMBO J. 19: 6475-88.

[35] Bardin AJ, Visintin R, Amon A (2000) A mechanism for coupling exit from mitosis to partitioning of the nucleus. Cell 102: 21-31.

[36] Bloecher A, Venturi GM, Tatchell K (2000) Anaphase spindle position is monitored by the BUB2 checkpoint. Nat Cell Biol 2: 556-558.

[37] Pereira G, Hofken T, Grindlay J, Manson C, Schiebel E (2000) The Bub2p spindle checkpoint links nuclear migration with mitotic exit. Mol Cell 6: 1-10.

[38] Pereira G, Schiebel E (2005) Kin4 kinase delays mitotic exit in response to spindle alignment defects. Mol Cell 19: 209-221.

[39] D'Aquino KE, Monje-Casas F, Paulson J, Reiser V, Charles GM, et al (2005) The protein kinase Kin4 inhibits exit from mitosis in response to spindle position defects. Mol Cell 19: 223-234.

[40] Geymonat M, Spanos A, Walzer PA, Johnston LH, Sedgwick SG (2003) In vitro regulation of budding yeast Bfa1/Bub2 GAP activity by Cdc5. J Biol Chem. 278: 14591-4.

[41] Hu F, Wang Y, Liu D, Li Y, Qin J, Elledge SJ (2001) Regulation of the Bub2/Bfa1 GAP complex by Cdc5 and cell cycle checkpoints. Cell. 107: 655-65.

[42] Maekawa H, Priest C, Lechner J, Pereira G, Schiebel E (2007) The yeast centrosome translates the positional information of the anaphase spindle into a cell cycle signal. J Cell Biol 179: 423-436.

[43] Stegmeier F, Amon A (2004) Closing Mitosis: The Functions of the Cdc14 Phosphatase and Its Regulation. Annu Rev Genet 38: 203-232.

[44] Chan LY, Amon A (2010) Spindle position is coordinated with cell-cycle progression through establishment of mitotic exit-activating and -inhibitory zones. Mol Cell. 39: 444-54.

[45] Caydasi AK, Pereira G (2009) Spindle alignment regulates the dynamic association of checkpoint proteins with yeast spindle pole bodies. Dev Cell 16: 146-156.

[46] Molk JN, Schuyler SC, Liu JY, Evans JG, Salmon ED, et al (2004) The differential roles of budding yeast Tem1p, Cdc15p, and Bub2p protein dynamics in mitotic exit. Mol Biol Cell 15: 1519-1532.

[47] Monje-Casas F, Amon A (2009) Cell polarity determinants establish asymmetry in MEN signaling. Dev Cell 16: 132-145.

[48] Pereira G, Tanaka TU, Nasmyth K, Schiebel E (2001) Modes of spindle pole body inheritance and segregation of the Bfa1p- Bub2p checkpoint protein complex. Embo J. 20: 6359-6370.

[49] Chan LY, Amon A (2009) The protein phosphatase 2A functions in the spindle position checkpoint by regulating the checkpoint kinase Kin4. Genes Dev 23: 1639–1649.

[50] Jill E, Falk L, Chan Y, Amon A (2011) Lte1 promotes mitotic exit by controlling the localization of the spindle position checkpoint kinase Kin4 PNAS 108: 12584-590.

[51] Caydasi AK, Kurtulmus B, Orrico MI, Hofmann A, Ibrahim B, et al (2010) Elm1 kinase activates the spindle position checkpoint kinase Kin4. J Cell Biol 190: 975-989.

[52] Moore JK, Chudalayandi P, Heil-Chapdelaine RA, Cooper JA (2010) The spindle position checkpoint is coordinated by the Elm1 kinase. J Cell Biol 191: 493-503.

[53] Edgington NP, Blacketer MJ, Bierwagen TA, Myers AM. (1999) Control of *Saccharomyces cerevisiae* filamentous growth by cyclin-dependent kinase Cdc28. Mol Cell Biol. 19:1369-80.

[54] Sreenivasan A, Kellogg D. (1999) The elm1 kinase functions in a mitotic signaling network in budding yeast. Mol Cell Biol. 19: 7983-94.

[55] Bouquin N, Barral Y, Courbeyrette R, Blondel M, Snyder M, Mann C. (2000) Regulation of cytokinesis by the Elm1 protein kinase in Saccharomyces cerevisiae. J Cell Sci. 113: 1435-45.

[56] Adams JA (2003) Activation loop phosphorylation and catalysis in protein kinases: is there functional evidence for the autoinhibitor model? Biochemistry. 42: 601-7.

[57] Bi E, Maddox P, Lew DJ, Salmon ED, McMillan JN, Yeh E, Pringle JR (1998) Involvement of an actomyosin contractile ring in *Saccharomyces cerevisiae* cytokinesis. J. Cell Biol. 142: 1301-12.

[58] Lippincott J, Li R (1998) Sequential assembly of myosin II, an IQGAP-like protein, and filamentous actin to a ring structure involved in budding yeast cytokinesis. J. Cell Biol. 140: 355-66.

[59] Yeong FM (2005) Severing all ties between mother and daughter: cell separation in budding yeast. Mol Microbiol. 55: 1325-31.

[60] Sanchez-Diaz A, Marchesi V, Murray S, Jones R, Pereira G, Edmondson R, Allen T, Labib K (2008) Inn1 couples contraction of the actomyosin ring to membrane ingression during cytokinesis in budding yeast. Nat Cell Biol. 4: 395-406.

[61] Nishihama R, Schreiter JH, Onishi M, Vallen EA, Hanna J, Moravcevic K, Lippincott MF, Han H, Lemmon MA, Pringle JR, Bi E (2009) Role of Inn1 and its interactions with Hof1 and Cyk3 in promoting cleavage furrow and septum formation in *S. cerevisiae*. J Cell Biol. 185: 995-1012.

[62] Lesage G, Bussey H (2006) Cell wall assembly in *Saccharomyces cerevisiae*. Microbiol Mol Biol Rev. 70: 317-43.

[63] Kuranda MJ, Robbins PW (1991) Chitinase is required for cell separation during growth of *Saccharomyces cerevisiae*. J Biol Chem. 266: 19758-67.

[64] Colman-Lerner A, Chin TE, Brent R (2001) Yeast Cbk1 and Mob2 activate daughter-specific genetic programs to induce asymmetric cell fates Cell. 107: 739-50.

[65] Bembenek J, Kang J, Kurischko C, Li B, Raab JR, Belanger KD, Luca FC, Yu H (2005) Crm1-mediated nuclear export of Cdc14 is required for the completion of cytokinesis in budding yeast. Cell Cycle. 4: 961-71.

[66] Frenz LM, Lee SE, Fesquet D, Johnston LH (2000) The budding yeast Dbf2 protein kinase localises to the centrosome and moves to the bud neck in late mitosis. J Cell Sci. 19: 3399-408.

[67] Luca FC, Mody M, Kurischko C, Roof DM, Giddings TH, Winey M (2001) *Saccharomyces cerevisiae* Mob1p is required for cytokinesis and mitotic exit. Mol Cell Biol. 21: 6972-83.

[68] Song S, Grenfell TZ, Garfield S, Erikson RL, Lee KS (2000) Essential function of the polo box of Cdc5 in subcellular localization and induction of cytokinetic structures. Mol Cell Biol. 20: 286-98.

[69] Xu S, Huang HK, Kaiser P, Latterich M, Hunter T (2000) Phosphorylation and spindle pole body localization of the Cdc15p mitotic regulatory protein kinase in budding yeast. Curr Biol. 10: 329-32.

[70] Lippincott J, Shannon KB, Shou W, Deshaies RJ, Li R (2001) The Tem1 small GTPase controls actomyosin and septin dynamics during cytokinesis. J Cell Sci. 114: 1379-86.

[71] Hwa Lim H, Yeong FM, Surana U (2003) Inactivation of mitotic kinase triggers translocation of MEN components to mother-daughter neck in yeast. Mol Biol Cell. 14: 4734-43.

[72] Korinek WS, Bi E, Epp JA, Wang L, Ho J, Chant J (2000) Cyk3, a novel SH3-domain protein, affects cytokinesis in yeast. Curr Biol. 10: 947-50.

[73] Ko N, Nishihama R, Tully GH, Ostapenko D, Solomon MJ, Morgan DO, Pringle JR (2007) Identification of yeast IQGAP (Iqg1p) as an anaphase-promoting-complex substrate and its role in actomyocin-ring-independent cytokinesis. Mol Biol Cell. 18: 5139-53.

[74] Jendretzki A, Ciklic I, Rodicio R, Schmitz HP, Heinisch JJ (2009) Cyk3 acts in actomyosin ring independent cytokinesis by recruiting Inn1 to the yeast bud neck. Mol Genet Genomics. 282: 437-51.

[75] Meitinger F, Petrova B, Lombardi IM, Bertazzi DT, Hub B, Zentgraf H, Pereira G (2010) Targeted localization of Inn1, Cyk3 and Chs2 by the mitotic-exit network regulates cytokinesis in budding yeast. J Cell Sci. 123: 1851-61.

[76] Frenz LM, Lee SE, Fesquet D, Johnston LH (2000) The budding yeast Dbf2 protein kinase localises to the centrosome and moves to the bud neck in late mitosis. J Cell Sci. 19: 3399-408.

[77] Yoshida S, Toh-e A (2001) Regulation of the localization of Dbf2 and Mob1 during cell division of Saccharomyces cerevisiae. Genes Genet. Syst. 76:141–47.

[78] Aspenström P (2009) Roles of F-BAR/PCH proteins in the regulation of membrane dynamics and actin reorganization. Int Rev Cell Mol Biol. 272: 1-31.

[79] Vallen EA, Caviston J, Bi E (2000) Roles of Hof1p, Bni1p, Bnr1p, and Myo1 in cytokinesis in Saccharomyces cerevisiae. Mol Biol Cell. 11: 593-611.

[80] Blondel M, Bach S, Bamps S, Dobbelaere J, Wiget P, Longaretti C, Barral Y, Meijer L, Peter M (2005) Degradation of Hof1 by SCF(Grr1) is important for actomyosin contraction during cytokinesis in yeast. EMBO J. 24: 1440-52.

[81] Meitinger F, Boehm ME, Hofmann A, Hub B, Zentgraf H, Lehmann WD, Pereira G (2011) Phosphorylation-dependent regulation of the F-BAR protein Hof1 during cytokinesis. Genes Dev. 25: 875-88.

[82] Martínez-Rucobo FW, Eckhardt-Strelau L, Terwissscha van Scheltinga AC (2009) Yeast chitin synthase 2 activity is modulated by proteolysis and phosphorylation. Biochem J. 417: 547-54.

[83] Teh EM, Chai CC, Yeong FM (2009) Retention of Chs2p in the ER requires N-terminal CDK1-phosphorylation sites. Cell Cycle 8: 2964–2974.

[84] VerPlank L, Li R (2005) Cell cycle-regulated trafficking of Chs2 controls actomyosin ring stability during cytokinesis. Mol Biol Cell. 16: 2529-43.

[85] Zhang G, Kashimshetty R, Ng KE, Tan HB, Yeong FM (2006) Exit from mitosis triggers Chs2p transport from the endoplasmic reticulum to mother-daughter neck via the secretory pathway in budding yeast. J Cell Biol. 174: 207–220.

[86] Chin CF, Bennett AM, Ma WK, Hall MC, Yeong FM (2012) Dependence of Chs2 ER export on dephosphorylation by cytoplasmic Cdc14 ensures that septum formation follows mitosis. Mol Biol Cell. 23: 45-58.

[87] Chuang JS, Schekman RW (1996) Differential trafficking and timed localization of two chitin synthase proteins, Chs2p and Chs3p. J Cell Biol.135: 597-610.

Histidine Kinases in Two-Component Systems

Chapter 7

Bacterial Two-Component Systems: Structures and Signaling Mechanisms

Shuishu Wang

Additional information is available at the end of the chapter

1. Introduction

Two-component systems (TCS) are ubiquitous among bacteria. They play essential roles in signaling events in bacteria, such as cell-cell communication, adaptation to environments, and pathogenesis in the case of pathogens. Due to their absence in humans and other mammals, TCS proteins are considered potential targets for developing new antibiotics. Most bacterial TCSs consist of two proteins, a sensor histidine kinase (HK) and a response regulator (RR). The HK senses specific signals, and that leads to activation of the kinase activity and autophosphorylation of a conserved histidine residue. The phosphoryl group is subsequently transferred to a cognate response regulator to activate its activities. Most RRs are transcription regulators and turn on or off gene transcriptions in response to the signals received by sensor HKs. Recent years have seen a rapid expansion of structural data of bacterial TCS proteins. In this chapter, I will review structures of HKs and RRs and discuss their structure-function relationship and their signaling mechanisms. The chapter contains three main sections: structures of histidine kinases, structures of response regulators, and structures of complexes between a histidine kinase and a response regulator. I will conclude the chapter with the implications of these structural and functional data on developments of new therapeutics against bacterial pathogens. Analysis of structures and reaction mechanisms will focus on the extracytosolic sensor HKs and OmpR/PhoB subfamily transcription regulator RRs.

2. Structures of histidine kinases

The prototypical histidine kinase is a homodimeric integral membrane protein (Figure 1). Each protomer has two transmembrane (TM) helices with the N-terminus in the cytosol. An extracytosolic sensor domain lies between the two TM helices. After the second TM helix is a HAMP domain, which is commonly found in Histidine kinases, Adenylyl cyclases, Methyl-accepting chemotaxis proteins, and Phosphatases [1], hence the name. The HAMP domain connects TM2 to the dimerization and histidine phosphorylation domain (often abbreviated as DHp). A catalytic and ATP-binding (CA) domain lies at the carboxyl terminus. The combination of DHp and CA is sometimes referred to as the kinase domain. The TM helices, HAMP domains, and DHp domains are all involved in homodimerization. The sensor domain is likely to dimerize in the context of the entire protein. Sensor domains share little sequence identity, as is expected from their diverse functions of sensing various signals. However, available structures of sensor domains fall into a few common structural folds, suggesting conserved signal sensing mechanisms. The HAMP, DHp, and CA domains are common modules of HKs and have well conserved structures and sequences, especially DHp and CA, whose sequences contain several conserved motifs. There is an absolutely conserved histidine residue that is phosphorylated, and the phosphoryl group is then donated to an RR, in response to signals sensed by the sensor domain. How the sensor domain regulates the kinase activities is still unknown, due to the lack of full-length structures of the transmembrane sensor HKs. However, a large accumulation of structures of isolated domains in recent years has started to shed light on possible molecular mechanisms of the signal transduction. In this section, I will summarize these structures and discuss their conformational changes that transmit signals from the sensor domain to the kinase domain.

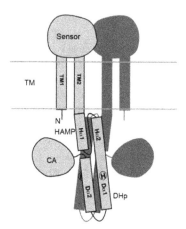

Figure 1. Schematic diagram of the modular structure of the prototypical sensor histidine kinase. At the N-terminus, a sensor domain lies between the two transmembrane (TM) helices. A HAMP domain follows TM2 with its second helix (Hα2) connected to helix Dα1 of the DHp domain. The catalytic and ATP-binding (CA) domain is at the C-terminus. The position of the phosphorylation site histidine is marked with an "H". The HAMP and DHp domains form homodimers.

2.1. Structures of sensor domains

Sensor domains of histidine kinases are located in cytosol, in membrane, or outside of cell membrane (extracytosolic). Currently, there is little structural information of the membrane-embedded sensor domains. The prototypical HK has an extracytosolic sensor domain that senses extracellular signals or conditions in the cell envelope. These sensor domains have highly diverse sequences. However, most of the known structures of extracytosolic sensor domains fall into three distinct structural folds, mixed αβ, all-helical, and β-sandwich. Unlike extracytosolic sensor domains, many cytosolic sensor domains can be annotated on the sequence level as PAS or GAF domains, which have related structural folds and are named from their occurrence in Period circadian, Aryl hydrocarbon receptor nuclear translocator, and Single-minded proteins (PAS) [2], or in cGMP-regulated cyclic nucleotide phosphodiesterases, Adenylate cyclases, and the bacterial transcriptional regulator FhlA (GAF) [3].

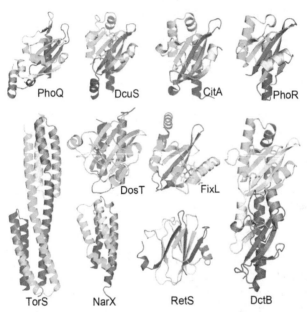

Figure 2. Structures of sensor domains. The structures are each colored in a rainbow spectrum from blue to red from the N- to C-terminus, respectively. The DosT (PDB code 2VZW, *M. tuberdulosis*) and FixL (1EW0, *R. meliloti*) sensor domains are located in the cell cytosol and belong to GAF and PAS domains, respectively; the remainders are extracytosolic domains. The PhoQ (3BQ8, *Escherichia coli*) and DcuS (3BY8, *E. coli*), and PhoR (3CWF, *Bacillus subtilis*), and CitA (2J80, *Klebsiella pneumoniae*) sensor domains have a PDC fold. DctB (3BY9, *Vibrio cholerae*) has two tandem PDC domains. The structures are shown in similar orientations with the N- and C-termini facing down for easy comparison, except DosT and FixL, which are shown with their central β-sheet having the same orientation as the PDC domains but their N- and C-termini facing other directions. The ligands and heme groups of DcuS, CitA, DctB, DosT, and FixL are shown as sticks. All of them are bound at the central β-sheet. NarX (3EZH, *E. coli*) has an all-helical sensor domain, while TorS (3I9Y, *Vibrio parahaemolyticus*) has double helical domains. The RetS (3JYB, *Pseudomonas aeruginosa*) sensor domain is currently the sole structural representative of β-sandwich fold extracytosolic sensors. The ribbon diagrams were prepared with PyMOL (Schrodinger, LLC).

DosS and DosT from *Mycobacterium tuberculosis* are soluble sensor HKs that have dual GAF domains located in the cytosol. The first GAF domain binds a heme group for sensing redox and hypoxia conditions of the cell cytoplasm to regulate the dormancy regulon *Dos*, which is believed by some researchers to allow *M. tuberculosis* to survive in a latent state for decades [4]. The structures of the N-terminal GAF domains of DosS [5] and DosT [6] reveal highly similar structures with a central five-stranded antiparallel β-sheet flanked on one side by the first and last helices and on the other side by some loops and short helices (Figure 2, DosT). The heme group is bound in the center of the β-sheet with the plane of the heme perpendicular to the β-sheet. The PAS domain of FixL from *Rhizobium meliloti* has the same topology of the central β-sheet, but has different α-helices (Figure 2) [7]. The heme binds at a different location of the sheet in a parallel manner.

The most common structural fold of the extracytosolic sensor domains is the mixed αβ fold, which has an identical topology to that of PAS domains [8]. The structures consist of a central 5-stranded antiparallel β-sheet, flanked by α-helices on both sides (Figure 2, top row). These extracytosolic sensor domains are slightly different in structure from the PAS and GAF domains. Their structures fall into a group by themselves and are named as PDC domains, referring to first three structures of this group, i.e. sensor domains of PhoQ, DcuS, and CitA [9]. Among the PDC domains, the structures are similar to each other, especially the N-terminal helix and the position of the C-terminus. The PDC domains have a long N-terminal helix, possibly continuous from the TM1 helix. The N- and C-termini of the domain are next to each other, and both are facing the same direction toward the membrane for connecting to the transmembrane helices. In some structures, there is a short helix at the C-terminus, possibly continuous to the TM2 helix.

Some HK sensor domains are all-helical, represented by those of NarX [10] and TorS [11] (Figure 2). The NarX sensor domain is an antiparallel four-helix bundle. Most of the helices have kinks or bends, suggesting mobility. TorS has two antiparallel four-helix bundles stacked along their bundle axes, with the second one inserted between the last two helices of the first bundle. Their mechanisms for signal sensing are different. NarX binds directly to nitrite and nitrate; while TorS interacts with a periplasmic binding protein TorT to detect trimethylamine-N-oxide.

The crystal structure of the sensor domain of RetS from *P. aeruginosa* reveals a β-sandwich fold (Figure 2) [12]. RetS works with two other HKs, LadS and GacS, to regulate genes for type III secretion system controlling acute infection and those for drug-resistant biofilm formation controlling chronic infection. The RetS sensor domain has two antiparallel β-sheets, with topologies β1-β3-β8-β5-β6 and β2-β9-β4-β7, stacking back-to-back. The β-sandwich fold structure resembles carbohydrate-binding modules.

2.2. Structure and function of HAMP domains

HAMP domains are widely occurring domains of prokaryotic transmembrane receptors. They follow the last TM helix and thus connect the extracytosolic sensory domain to the cytosolic signaling domains. The sequences of HAMP domains have heptad repeats, in which hydrophobic residues occupy positions a and d (Figure 3). The structures of HAMP domains are dimers with each protomer of ~50 residues forming two helices (referred to as Hα1 and Hα2) connected by a linker of ~14 residue. The four helices from two protomers associate to form a parallel four-helix bundle. The linker residues have an extended structure that spans the length of the four-helix bundle [13, 14]. Other than the hydrophobic side chains of the heptad repeats and a conserved glycine residue at the beginning of the linker, HAMP domains do not have strong sequence conservation. Nevertheless, the domains can be exchanged, yielding hybrid proteins that are still functional [15, 16], which suggests that HAMP domains share a conserved mechanism for propagating signals.

The HAMP domain of Af1503, a hypothetical receptor from the archaeon *Archaeoglobus fulgidus* [14] is the first HAMP structure determined (Figure 3). The packing of the 4 helices is different from the expected conventional knobs-into-holes geometry, in which a residue from one helix (knob) packs into a space surrounded by four side chains of the facing helix (hole). Instead, the helical packing of the Af1503 HAMP domain is described as complementary x-da, in which the x- and da-layers alternate along the helices to form mixed x-da layers. In the x-layer, side chains point straight at the central supercoil axis; and in the da-layer, side chains point side ways to form an interacting ring of residues enclosing a central cavity. This non-canonical packing geometry is stabilized by the small side chain of alanine at residue 291 [17]. Mutations of A291 with amino acids of larger side chain convert the helical packing to the knobs-into-holes conformation through a concerted axial rotation of all four helices of ~26° (Figure 3). It was hence proposed that signals propagate through the TM helices and HAMP by helix rotation around an axis perpendicular to the membrane, a so-called gearbox model [18].

The crystal structure of a tri-HAMP unit from the *P. aeruginosa* soluble receptor Aer2 shows that the three HAMP domains each form a parallel four-helix bundle similar to HAMP of Af1503 (Figure 3) [13]. This confirms that the parallel four-helix bundle is likely universal among HAMP domains. The top HAMP domain (HAMP1) is separated from HAMP2 by a helical insert that connects Hα2 of HAMP1 to Hα1 of HAMP2 by a slightly kinked helical insert. HAMP2 and HAMP3 are contiguous and form a concatenated di-HAMP structure. There are two distinct packing configurations among the three HAMP domains. HAMP1 and HAMP3 are similar to each other and resemble that of Af1503. The middle HAMP2 domain, however, has a unique helical packing, in which there is an offset of the helical register between α1 and α2 of half a helical turn (~2-3 Å). This results in side chain staggering of the buried hydrophobic core between the two helices of the same polypeptide chain. The two α1 helices coil around each other, so do the two α2 helices; but the two coiled-coils have limited interactions with each other. The unique packing of HAMP2 leads to a slightly different possible mechanism of signaling propagation by the HAMP domain: a combination of vertical movement and helical rotation, also known as screw-like motion.

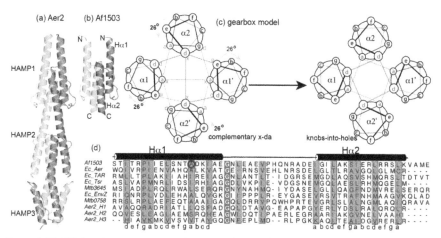

Figure 3. HAMP domain structures, a gearbox model for signal propagation, and sequences. The crystal structure of a tri-HAMP domain from Aer2 (Pdb code 3LNR) shows that all three HAMP domains have a four-helix bundle structure (a), which is similar to that of the HAMP domain of Af1503 (2L7H) (b). Each four-helix bundle is a homodimer of a parallel two-helix coiled-coil of helix Hα1 and Hα2. Af1503 has a helical packing of complementary x-da, which can be converted to the canonical knobs-into-holes supercoil by a 26 º rotation of each helix (c). (d) Sequence alignment of HK HAMP domains shows heptad repeats with hydrophobic side chains at positions a and d. Conserved hydrophobic residues are highlighted in red. There is a highly conserved Gly at the beginning of the linker, highlighted in blue. Residue A291 in Af1503 is circled. Aer2_H1, Aer2_H2, and Aer2_H3 in (d) are sequences for Aer2 HAMP1, HAMP2, and HAMP3, respectively.

2.3. The dimerization and phosphorylation domain

The dimerization and phosphorylation domain is a conserved domain of histidine kinases. This domain is often abbreviated as DHp for dimerization and histidine phosphotransfer. It is also referred to as dimerization domain or histidine kinase domain A. The sequence of DHp contains the H box that is part of the signature motifs defining histidine kinases (Figure 4). The histidine residue in the H box motif is absolutely conserved and is the site for autophosphorylation and subsequent transfer of the phosphoryl group to downstream proteins. This domain participates in dimerization of the histidine kinase. DHp is located in the cytosol and directly follows the HAMP domain, and it is immediately followed by the CA domain in the prototypical histidine kinases (Figure 1).

The structure of the DHp domain consists of two α-helices (referred to as Dα1 and Dα2), that form an antiparallel coiled-coil (also referred to as helical hairpin), and two protomers associate to form a four-helix bundle (Figure 5). The conserved histidine residue is on helix Dα1. Similar to HAMP domains, sequences of the DHp domains have heptad repeats. Several structures of DHp have been determined. The NMR structure of the *E. coli* EnvZ DHp shows a four-helix bundle that is relatively straight [19]. The phosphorylation site H243 is solvent exposed and located in the middle of helix Dα1, which is more mobile than

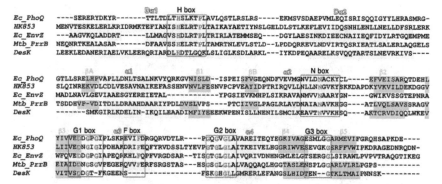

Figure 4. Sequences of the DHp and CA domains of HKs. Conserved sequence motifs H, N, G1, F, G2, and G3 boxes are in magenta boxes. Helices are highlighted in yellow, and β-strands in green.

helix Dα2, especially around the conserved H-box residues. The intra-subunit interface is mostly hydrophobic; the dimer interface, however, contains two acidic clusters. A related phosphotransferase SpoOB from *B. subtilis* has a similar dimerization domain of four-helix bundle [20]. The active site residue H30 is in the middle of helix Dα1, with its side chain protruding to the surface, similar to that of the EnvZ DHp domain.

The structure of the entire cytoplasmic portion of HK853, a putative sensor HK from *Thermotoga maritima*, shows that the DHp domain forms a four-helix bundle slightly different from that of EnvZ [21]. The HK853 four-helix bundle has a larger twist angle of ~25°, compared to that of EnvZ, which is unusually straight and parallel. The connections between hairpin helices have different topologies between the two structures. EnvZ is unique in this respect as all other known DHp structures have the same topology as HK853 (Figure 5). HK853 does not have a HAMP domain but has a linker of 22 residues between TM2 and DHp. This linker region forms a helix, possibly continuous from TM2 to Dα1. The two helices of the dimer associate to form a left-handed coiled-coil on top of the four-helix bundle. There is a kink in Dα1 induced by P265, which is near the phosphorylation site H260 and is well conserved among HKs. This kink is common among all DHp structures, even those without a proline, such as SpoOB and DesK (Figure 5).

2.4. ATP-binding domain

In prototypical sensor histidine kinases, the ATP-binding domain is located at the carboxyl terminus of the polypeptide (Figure 1). The ATP-binding domain is often abbreviated as CA for catalytic and ATP-binding. This domain binds ATP, which then donates its γ-phosphate group to the conserved histidine residue on the DHp domain. Sequences of the CA domains have well conserved N, G1, F, G2, and G3 box sequence motifs (Figure 4). Together with the H box at the DHp domain, they are the markers to annotate histidine kinases from DNA sequences.

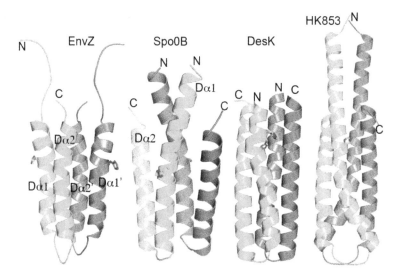

Figure 5. Structures of the DHp domains. All structures show an antiparallel helical bundle. The phosphorylation site histidine is shown as sticks. This histidine residue is located in the middle of Dα1. There is a kink of the helix right below the histidine residue, produced by a strongly conserved proline residue in EnvZ (PDB ID 1JOY) and HK853 (2C2A). DesK (3EHF) and Spo0B (1IXM) do not have this proline residue but still have a kink right below the histidine residue. Structures of DHp from Spo0B, DesK, and HK853 are a part of larger structures containing a C-terminal domain.

The structures of the CA domains are well conserved, as expected from their highly conserved sequences. The structures have an αβ sandwich fold containing two layers, a layer of mixed 5-stranded β-sheet and a layer of three α-helices (Figure 6). The CA domain of DesK from *B. subtilis* has the shortest sequence and represents the minimal core structure for this domain [22]. Most of other structures have additional α-helices, β-strands, and a longer ATP-lid (the loop covering the ATP-binding site). The ATP-binding site in DesK is much shallower due to the lack of additional structural elements and a shorter ATP-lid. Interestingly, an intact ATP molecule binds in the ATP-binding site with full occupancy, even though the phosphate groups are partially exposed to the protein surface. It is likely that in the crystal, the protein is in a conformation that shields any nucleophilic water from getting access to the phosphates, thereby protecting the ATP from hydrolysis during crystal growth. In this case, phosphotransfer will occur only in the presence of the DHp domain with the conserved histidine functioning as a nucleophile to transfer the γ-phosphate to the histidine.

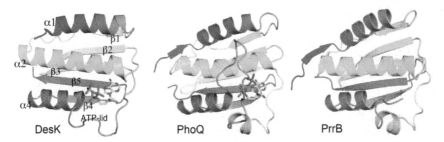

Figure 6. Ribbon diagrams of the ATP-binding domains. Each domain is colored in a rainbow spectrum with blue for the N-terminus and red for the C-terminus. Bound ATP in DesK (PDB code 3EHG) and AMPPNP in PhoQ (1ID0) are shown as sticks. The structure of PrrB (1YS3) from *M. tuberculosis* does not have a ligand bound, and part of the ATP-lid is disordered. Helices and β-strands are labeled on the structure of DesK CA. The short helix α3 in the ATP-lid is missing in DesK.

The ATP-binding site is at one end of the domain (Figure 6) and involves both absolutely conserved and partially conserved residues from the N, G1, F, G2, and G3 boxes (Figure 4). Residues from the G1 and G3 boxes are involved in binding the adenosine moiety; those from the N and G2 boxes contact both the adenosine moiety and the triphosphates-Mg^{2+} moiety. The structure of *E. coli* PhoQ in complex with AMPPNP reveals detailed binding interactions of the conserved boxes with ATP [23]. In the G1 box, the conserved Asp has a direct H-bond to the amino group and a water-mediated H-bond to N1 of adenine. In the N box, the side chains of conserved N389 and partially conserved N385 interact with the α-phosphate; the side chain oxygen of conserved N389 has a water-mediated H-bond to N1 of adenine; the partially conserved K392 and Y393 interact with the γ-phosphate through their side chains. The side chain of Y393 stacks with the adenine ring, while that of I420 from G1 box flanks the other face of the adenine ring. The side chains of partially conserved N385 (N box) and Q442 (G2 box) coordinate the divalent cation, which binds to the triphosphate group. DesK has similar binding interactions with the adenosine moiety, but slightly different binding interactions with the phosphate groups, although both proteins involve residues of the N and G2 boxes for binding the phosphate groups.

A long loop covering the ATP-binding site is called ATP-lid, which spans from the F box to the beginning of helix α4 and plays an important role in binding ATP. The ATP-lid is highly mobile; it has several glycine residues, including the conserved G2 box that contains three glycine residues (Figure 4). The bound ATP has extensive contacts with the ATP-lid residues, and thus ATP binding induces the closure of the ATP-lid. In the absence of ATP, this loop is partially disordered in crystal structures. Even in the presence of ATP, the ATP-lid shows high flexibility, indicated by high B-factors or in some cases is partially disordered in crystal structures [21-24]. The flexibility of the ATP-lid is important not only for binding ATP, but also for interacting with the DHp domain for the phosphotransfer reactions. The ATP-binding domain must adopt several positions relative to the DHp domain because the HK functions as an autokinase, phosphotransferase, or phosphatase in response to the environmental stimulates received. The flexibility of the ATP-lid allows the CA domain to bind to different regions of DHp depending on the conformation of the DHp domain and the status of the ATP-binding site of the CA domain.

2.5. Signal transduction mechanism from sensor domain to the kinase domain

Signals flow from the sensor domain to the kinase domain by conformational changes throughout the entire protein. Environmental cues received by the sensor domain cause a conformational change in the sensor domain. That conformational change is transmitted through the transmembrane helices and the HAMP domain to the kinase domain. Most of isolated sensor domains are monomers in solution. However, they are expected to form dimers in the context of the whole protein because the DHp and HAMP domains are homodimers. Therefore, ligand-induced dimerization cannot be a mechanism of signal transduction through the cell membrane. Instead, there are structural changes of the sensor domains upon ligand binding that could trigger rotation or piston-like translation movements of the TM helices [9, 25].

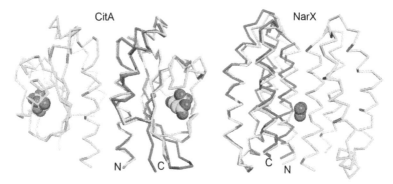

Figure 7. Piston model of ligand-induced conformational changes in the sensor domain. In both CitA (left panel, Pdb codes 2J80 and 2V9A) and NarX (right panel, 3EZH and 3EZI), ligand-bound structures are dimers in the crystal (green molecules), while ligand free forms are monomers (magenta color). Ligands are represented as space-filling models.

Comparison of structures of the sensor domain of CitA with and without citrate bound suggests a piston-type transmembrane signaling [8]. The sensor domain of CitA has a central 5-stranded anti-parallel β-sheet flanked by α-helices on both sides (Figure 2). The β-sheet curls up to one side where citrate binds. There are two loops, referred to as major loop and minor loop, which cover the citrate-binding site. The major loop is disordered in the citrate-free structure and becomes ordered in the presence of citrate in the binding site. Binding of citrate also induces more bent of the central β-sheet. This results in lifting of the C-terminus of the domain relative to the N-terminal helix (Figure 7). The C-terminus of the domain connects to the second transmembrane helix (TM2). Therefore, binding of citrate induces a piston-type movement of TM2 relative to TM1. Similar piston-type movement has been proposed for NarX from *E. coli*, whose sensor domain is a four-helix bundle that forms a dimer when binding nitrate (Figure 7) [10]. Binding of nitrate induces the movement of the N-terminal helix toward the cell membrane, relative to the C-terminal helix. Since the N- and C-terminal helices are expected to connect to TM1 and TM2, respectively, the relative movement of the two helices would push down TM1 or lift up TM2. This movement of helices could also translate into rotations of the TM helices, if winding or unwinding of helical turns occurs.

A rotation or screw-like motion of TM2 has been proposed as the signal transduction mechanism from the sensory rhodopsin into the cytosol for HtrII [26]. Based on the structure of HtrII, the transmembrane helices of prototypical HKs are expected to form a four-helix bundle [27]. A rotation or screw-like motion of the TM helices for propagating signal through membrane is compatible with that proposed for the HAMP domain, which is often found between TM2 and the DHp domain. Structural and biochemical studies suggest that HAMP domains are not rigid structures. They can alternate between different structures through helical rotation [14] or a shift in register of coiled-coils and rotation [13]. Mutation of an alanine residue of the HAMP domain of Af1503 into larger side chain residues changes the helical packing gradually from complementary x-da to the canonical knobs-into-holes mode [17]. Through these conformational changes, the HAMP domain passes the signals from the sensor domain to the downstream kinase domain.

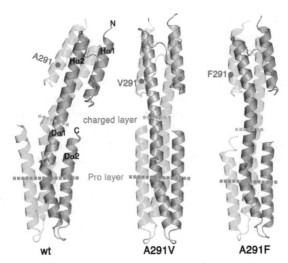

Figure 8. Crystal structures of chimeras of Af1503 HAMP fused to DHp of EnvZ. Mutations at A291 of HAMP cause conformational changes in the HAMP helical packing, which propagate to the DHp domain. The mutation sites are labeled with a red dot. Approximate positions of the charged layer and proline layer are marked. PDB codes for the structural models are 3ZRX (wt), 3ZRW (A291V), and 3ZRV (A291F).

Structures of chimeras of the Af1503 HAMP fused with the DHp domain of EnvZ show that conformational changes in the HAMP domain can trigger changes in the structure of the downstream DHp domain [18]. Figure 8 shows three crystal structures of the chimeras. The HAMP domains preserve the same conformations as their corresponding isolated domains. Mutations of A291 into amino acids with larger hydrophobic side chains change the helical packing conformation. These conformational changes in the HAMP domain propagate to the DHp domain through the connecting helices. The helices linking the HAMP domain to the DHp domain do not have a strong hydrophobic core as in the four-helix bundles but are relatively flexible, showing variations among the structures with different mutations. Conformational variation is the most pronounced at the junction between the DHp four-helix bundle and the two-helix coiled-coil where there are charged residues (charged layer, Figure 8). The induced rotational movements in helix Dα1 gradually diminish at the conserved proline, which produces a kink in the helix. Helix Dα2 also changes conformation responding to changes of Dα1, and as a result the bundle radius differs above the proline layer among variant structures.

The structural changes in DHp induced by the HAMP domain are likely to alter the activities by regulating interactions between the DHp and CA domains. Despite the structural changes triggered by varying conformations in HAMP, the phosphorylation site histidine remains little perturbed, and thus altering accessibility of the conserved histidine is unlikely to be the mechanism of signal transmission. The lower half of the DHp helical bundle (below the proline layer) remains invariant among the structures. This region contains many residues important for binding the cognate RR and contributes the majority of the interface for RR binding in the HK-RR complex structures [19, 28]. Therefore, blocking direct binding of RR is unlikely to be a mechanism for signaling either. However, the region of DHp most variable among the structures (charged layer) plays an important role in interacting with the CA domain. In the structure of HK853-RR468 complex, residues F428 and L444 of the ATP-lid interact with an exposed hydrophobic core of DHp near the charged layer, and this interaction locks the ATP-binding domain in a kinase-inactive state [28]. It is likely that conformational changes triggered by signals transmitted from the HAMP domain either promote this DHp-CA interaction (phosphatase state) or release the CA domain for it to phosphorylate the conserved histidine (kinase state).

Several structures of the entire kinase domain (DHp + CA) are available and give insights into the interactions between the two sub-domains. The crystal structure of the entire cytoplasmic portion of HK853 from *T. maritima* shows a conformational state that is ready for the phosphotransfer reaction [21]. The HK853 cytoplasmic domain forms a dimer through the DHp domain (Figure 9a). The CA domain is connected to the DHp domain through a short linker. The crystal structure contains an ADPβN, from hydrolyzed AMPPNP, in the CA domain. The ATP-lid is flexible and partially disordered, despite the presence of the ADP analog in the ATP-binding site. The H260 side chain is fully exposed to the surface of the protein. There is a sulfate ion in the crystal structure bound next to the Nε atom of H260, mimicking the phosphorylated histidine. Contacts between DHp and CA are exclusively within one polypeptide chain and involve mostly conserved hydrophobic residues. The importance of the interface residues in the HK function is confirmed by site-directed mutagenesis studies [21]. In the DHp domain, the interface involves both α helices

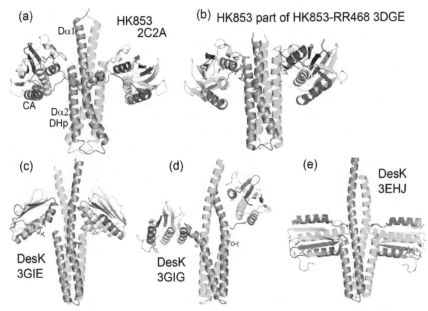

Figure 9. Structures of the entire kinase domain containing the DHp and CA subdomains. DHp domains are colored in orange for one subunit and green for the other. CA domains are colored in a rainbow spectrum. ATP analogs and the essential histidine side chains are shown as sticks except in (a) where the H260 side chains and a sulfate are shown as space-filling models. The hinge pivot residue G243 is shown as a green sphere in (c).

near the charged layer (Figures 8 and 9), and in the CA domain it involves the ATP-lid, helix α4, and residues at N-terminus of α2. The interactions are relatively weak, and thus the domain interface is dynamic, allowing the CA domain to move to other locations on DHp. The signals from the sensor domain control the interfacial stability through conformational changes of the helical bundle. The DHp-CA interface is destabilized for the kinase activity and stabilized for the phosphatase activity.

The crystal structure of HK853 complexed with RR468 reveals another binding interface between DHp and CA that is created by conformational changes in the helical packing [28]. In this structure, the CA domain moves closer to H260 of the same subunit (Figure 9b). This suggests that the autophosphorylation occurs in *cis*, although in the crystal structure, the ATP-lid interacts with the bound RR and blocks the access of the phosphorylation site histidine to ATP. The structure has a sulfate near each phosphoacceptor aspartate of RR, mimicking the phosphorylated RR. Therefore, the structure represents a conformational state poised for dephosphorylation of the RR. This phosphatase-competent state has a stronger domain interface between DHp and CA. A rotation of the helix Dα1 N-terminal coiled-coil region and a concerted inverse twist of Dα2, as in the proposed gearbox model for transmitting signals through HAMP domains (Figure 3), expose parts of the hydrophobic core of DHp near the charged layer. This allows residues F428 and L444 of ATP-lid to interact with the exposed hydrophobic pocket of DHp of the same subunit, and thus allows the CA domain to move to a new binding site.

Structural plasticity of the HK proteins is essential to allow conformational changes in order to transmit signals received at the sensor domain to the kinase domain and switch its enzymatic activities. This structural plasticity is well demonstrated by the structures of the cytoplasmic domain of the *B. substilis* DesK, a thermo-sensing histidine kinase [24]. The DHp domain is highly mobile, adopting different conformations to allow the CA domain to bind at different positions in different functional states. The structures reveal three distinct conformational states with different helical packing of the DHp domain and the relative position of the CA domain. One conformational state represents the autokinase-competent state. This conformational state is characterized by the lack of specific interactions between the DHp and CA domains, allowing the CA domain to move to a position suitable for autophosphorylation. A single rigid-body rotation of the CA domain along the interdomain hinge residue G243 can bring the ATP γ-phosphate close to the side chain of the phosphorylation site histidine of the other protomer (Figure 9c). In addition, the contact surfaces are complementary to each other in both contour and charges. The structures are consistent with a *trans* phosphorylation mechanism. That the autokinase-competent state lacks specific interdomain interactions is consistent with observations that isolated DHp domains can be phosphorylated in the presence of isolated CA domains. A second conformational state is poised for the phosphotransfer reaction, represented by structures of a phosphorylated DesK and a phosphomimetic H188E mutant. Both structures show a similar asymmetric homodimer. One of the CA domains interacts with both Dα1 helices at the coiled-coil region. This interaction precludes the binding of the other CA domain, which makes no contacts with the DHp domain in the phosphorylated DesK structure (Figure 9d), but is completely disordered in the H188E-ADP structure. The helical bending at the phosphorylation site histidine is more pronounced partially due to the negatively charged phosphohistidine associating with conserved basic residues of the opposite protomer. The third conformational state is a phosphatase-competent state, inferred from the structures of the H188V mutant. This mutation abolishes the kinase activity but retains the wild-type level phosphatase activity. The V188 side chain is buried in the hydrophobic core of the four-helix bundle (Figure 9e). This removes the kink in Dα1 and allows a tight packing rigid helical bundle. The changes in the structure of DHp also expose a patch of hydrophobic surface on Dα1 near the charged layer for a tight association with the CA domain. The CA domain interacts with the same surface of DHp as seen in the structures of HK853 [21] and KinB [29].

3. Structures of response regulators

Response regulators are simpler in structure than histidine kinases. The prototypical RR has two domains: a receiver domain that accepts a phosphoryl group from a cognate HK and an effector domain that generates the outputs of the signaling events. The structure and sequence of receiver domains are well conserved. In contrast, effector domains are variable, especially at the sequence level, reflecting their diverse output functions. The majority of RRs are transcription regulators with their effector domains as DNA-binding domains. A significant fraction of RRs have effector domains as enzymes. Many others have effector domains for binding RNA, ligands, or proteins to regulate bacterial cellular process at post-transcriptional and post-translational levels. There are also single domain RRs that have only the receiver domains such as CheY and Spo0F. RRs have been studied extensively for decades, and an extensive amount of structural data is available, as well as many well-written reviews [30]. I will focus my efforts on transcription regulator RRs and will give a summary on how phosphorylation of the receiver domain controls the activities of the effector domain with respect to their structure-function relationship.

3.1. Receiver domain structures

Receiver domains have well-conserved sequences and structures, suggesting a common signaling mechanism by two component systems to pass signals from HKs to their cognate RRs. The receiver domain is also known as the phosphorylation domain or regulatory domain. There is an invariable aspartate residue in the receiver domain that accepts a phosphoryl group from a cognate HK. Phosphorylation results in conformational changes in the domain, which is transmitted to the effector domain to regulate its activities. Receiver domains have a conserved $(\beta\alpha)_5$ fold (Figures 10 and 11), in which alternating β-strands and α-helices in the sequence fold into a central five-stranded parallel β-sheet surrounded by helices α1 and α5 on one side and helices α2, α3, and α4 on the other side. The β-sheet consists of mainly hydrophobic side chains that form a hydrophobic core. The α-helices are amphipathic and pack against the central β-sheet with their hydrophobic face. Most of the conserved residues are at the C-terminal ends of strands β1, β3, and β4. Helix α1 is involved in binding to the DHp domain of HK [28]. Although the structure and position of this helix is well conserved among RR structures, the amino acid sequence conservation is limited to the hydrophobic residues involved in packing against the hydrophobic core of the β-sheet (Figure 10). This helix is likely to play an important role in the specificity of HK-RR pairs.

The phosphorylation site is located at an acidic pocket near the C-termini of strands β1 and β3, where the conserved acidic side chains are clustered together. These acidic side chains are involved in binding a divalent cation, Mg^{2+} or Mn^{2+}, that is essential for the phosphorylation reaction [31]. An absolutely conserved aspartate at the end of the strand β3 is the phosphoacceptor. The loop β3-α3 forms a conserved structure that is held together by conserved residues: a Pro at 4 residues after the phosphorylation site aspartate and a Gly, which is at 4 residues after the Pro in sequence and the first residue of helix α3 (Figure 10).

The α4-β5-α5 face of the receiver domain has been proposed to be important for the function of RRs [32, 33]. Many RRs in the OmpR/PhoB subfamily are thought to be dimers in their active form. Structures of several RR receiver domains are found to form dimers through the α4-β5-α5 face when activated. Exposed side chains on this surface are primarily hydrophilic, suggesting that any interactions through this α4-β5-α5 face are likely to be dynamic. Comparison among receiver domain structures reveals that this face is considerably

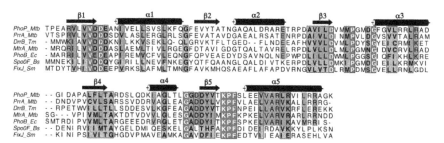

Figure 10. Sequence alignment of the response regulator receiver domains. The top five sequences, PhoP, PrrA and MtrA from *M. tuberculosis*, DrrB from *T. maritima*, and PhoB from *E. coli*, belong to the OmpR/PhoB subfamily. SpoOF from *B. subtilis* is a single domain RR. FixJ of *S. Meliloti* is in the NarL/FixJ subfamily. Identical residues for all sequences are shaded in blue. Highly conserved residues in the OmpR/PhoB subfamily are shaded in red. Moderately conserved residues across all sequences are shaded in gray.

variable. The sequence of helix α5 is well conserved among OmpR/PhoB subfamily RRs (Figure 10). Yet the length of this helix varies among known structures [34]. In the structure of PrrA from *M. tuberculosis* this helix has different lengths between two molecules in the crystal asymmetric unit [35]. Helix α4 and its flanking loops also have variable structures among RRs. Flexibility of the α4-β5-α5 elements is likely to be important for phosphorylation regulation of RR activities.

Activation of RRs by phosphorylation induces structural changes in the α4-β5-α5 face. This is accomplished by movements of two key residues, a conserved Thr/Ser at the C-terminus of β4 and a moderately conserved Tyr/Phe in the middle of β5 (Figure 11). These residues are referred to as switch residues. In the unphosphorylated RR, the side chain of Thr/Ser is oriented away from the phosphoacceptor, and that of Tyr/Phe extends outward toward the surface of the α4-β5-α5 face. Structures of activated receiver domains reveal that these side chains move in response to phosphorylation [36, 37]. All activated receiver domains of the OmpR/PhoB subfamily form a symmetric dimer through the same interface involving α4-β5-α5. Phosphorylation of the phosphoacceptor aspartate residue allows the side chain of the Thr/Ser residue to move closer to the phosphate group to make a favorable hydrogen bond. Repositioning the Thr/Ser side chain shifts the β4-α4 loop and helix α4, and makes the inward conformation of the Tyr/Phe residue more energetically favorable. Another conserved residue, a Lys in the β5-α5 loop, forms a salt bridge to the phosphoryl group and brings some modest shift in the β5-α5 loop [31]. This Lys side chain has a salt bridge to the side chain of phosphoacceptor aspartate in the unphosphorylated structure PhoP from *M. tuberculosis* [34]. It plays an important role in stabilizing the active site structure, but probably has only minor effects on structural switch between activated and non-activated RRs. Phosphorylation of NarL/FixJ subfamily RRs produces similar conformational changes in switch residues and results in dimerization through the α4-β5-α5 face, although the dimer interface are not identical to that of OmpR/PhoB RRs (Figure 11d). The difference in dimer interface might be due to lack of divalent cation in the crystal structure of FixJ [38].

Activation by BeF$_3$ and Mg^{2+} did not induce symmetric dimer formation of single domain response regulator RR468 [28]. However, the switch residues are in the active conformation as observed for all other RRs, suggesting that conformational changes of switch residues is the common mechanism for all RRs.

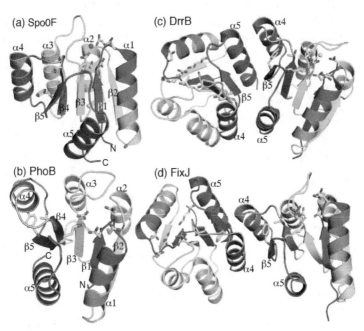

Figure 11. Structures of the receiver domains of response regulators. Each individual domain is colored from N- to C- terminus in a rainbow spectrum. Phosphorylation site residues and switch residues are shown as sticks. SpoOF (PDB code 1PEY) from *B. subtilis* binds a Mn^{2+} ion, shown in magenta. Both PhoB (1B00) from *E. coli* and SpoOB are non-activated. DrrB (3NNS) from *T. maritima* is activated with binding of BeF$_3$ and Mg^{2+}, shown as sticks and a green sphere, respectively. FixJ (1D5W) from *S. Meliloti* is activated by phosphorylation, but without divalent cations.

3.2. Structures of the effector domains

Unlike the receiver domains, the effector domains are very diverse, reflecting the diversity of cellular processes that are controlled by two-component systems. Effector domains are also referred to as output domains. They can be DNA-binding, RNA-binding, ligand-binding, or transporter output domains, or enzymes [30]. Except in archaea, which has almost 50% of RRs containing only the receiver domain, a great majority of bacterial RRs are transcription regulators that contain a C-terminal effector domain as a DNA-binding domain.

DNA-binding effector domains are grouped into several subfamilies based on predicted domain structures from amino acid sequences [30]. The largest subfamily is the OmpR/PhoB group, whose structure has a winged helix-turn-helix (wHTH) motif. The second largest subfamily is the NarL/FixJ group, which has a helix-turn-helix (HTH) DNA-binding motif. Another subfamily LytR/AgrA is also fairly common. A structure of a member of this subfamily, LytTR from *S. aureus*, shows that the DNA-binding domain consists mostly of β-strands [39]. The loops between the β-strands interact with DNA bases with side chains inserting into both major and minor grooves of DNA. Other subfamilies of DNA-binding response regulators are less common and are present only in certain types of bacterial species.

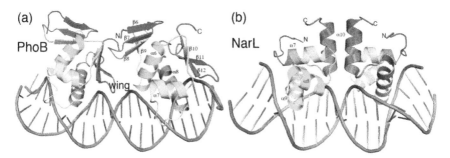

Figure 12. Structures of response regulator effector domains binding to DNA. (a) PhoB (1GXP) of *E. coli* binds to direct repeat Pho Box DNA as a tandem dimer. The PhoB effector domain has a winged HTH fold. (b) NarL (1JE8) of *E. coli* recognizes palindromic DNA sequences and forms a symmetric dimer through association of helix α10 on binding DNA.

The DNA-binding domain of the OmpR/PhoB subfamily has a conserved winged helix-turn-helix fold (Figure 12a). The structure starts with an N-terminal four-stranded anti-parallel β-sheet (β6 to β9), followed by three α-helices (α6 to α8) in the middle, and a β-hairpin (β11 and β12) at the C-terminus. A short strand (β10) between α6 and α7 assembles with the C-terminal β-hairpin to form a three-stranded anti-parallel β-sheet. The two β-sheets sandwich the three α-helices in the middle. The β-hairpin is called the wing, and α7, α8 and loop α7-α8 together constitute the helix-turn-helix motif. Helix α8 is the sequence recognition helix, which inserts into the major groove of DNA [40]. The β-hairpin (wing) binds in the minor groove of DNA. Loop α7-α8 is termed transaction loop because it is essential for transcription activation.

The C-terminal DNA-binding domain of the NarL/FixJ subfamily is a compact bundle of 4 α-helices (Figure 12). The second and third helices (α8 and α9) form a helix-turn-helix motif that interacts with the major groove of DNA [41, 42]. Helix α9 is the recognition helix that inserts into major groove of DNA with side chains interacting with DNA bases. The proteins bind DNA as a symmetric dimer to recognize palindromic DNA repeats. Helix α10 and the loop α7-α8 form the protein dimer interface in the protein-DNA complex. Isolated DNA-binding domains, however, are found to be monomeric in solution when not binding to DNA [42].

3.3. Structures of full-length response regulators and regulation mechanism

Full-length structures are available for the OmpR/PhoB and NarL/FixJ subfamilies of response regulators. Even though most RRs have only two domains, full-length structures prove to be difficult to obtain, especially for the OmpR/PhoB subfamily. This is most likely due to the dynamic nature of interactions between the receiver and effector domains. The domains are individually folded, and their associations are transient if they do associate with each other. The dynamic and transient nature of domain interactions is important for the regulation of DNA-binding activities. Signals transmitted from the HK through phosphorylation of the receiver domains must be able to relay to the DNA-binding domain by conformational changes of the receiver domain. Phosphorylation promotes dimerization of the receiver domain and in some cases releases the blockage of the DNA-binding element of the receiver domain. So far, all available structures of full-length RRs are non-activated and without DNA bound.

The full-length NarL structure in its unphosphorylated form reveals that the receiver domain is associated with the DNA-binding domain through an interface involving the DNA-binding elements [43]. The linker between the two domains forms a helix ($\alpha 6$) followed by a flexible loop that is disordered in the crystal structure (Figure 13). Helix $\alpha 6$ packs against $\alpha 4$ of the receiver domain and $\alpha 10$ of the effector domain. The domain interface also involves helices $\alpha 8$ and $\alpha 9$, locking the DNA-sequence-recognition helix $\alpha 9$ in a manner incapable of inserting into the DNA major groove. Phosphorylation is expected to dissociate the effector domain from the receiver domain, thus allowing it to bind DNA. Unphosphorylated NarL does not bind DNA in solution. NarL binds to inverted-repeat DNA sequences as a symmetric dimer, whose interface involves helix $\alpha 10$ and loop $\alpha 7$-$\alpha 8$ (Figure 12). Phosphorylation may also promote dimerization of the receiver domain and thus strengthens the NarL dimer.

Full-length structures of unphosphorylated StyR and DosR, which are members of the NarL/FixJ subfamily RRs, are also known (Figure 13). The receiver domain of StyR from *Pseudomonas fluorescens* has a conserved $(\beta\alpha)_5$ fold [44]. Its DNA-binding domain has the same fold as that of NarL. However, the interdomain linker forms a long helix continuous from $\alpha 5$ of the receiver domain that separates the N- and C-terminal domains by more than 16 Å, exposing the helix-turn-helix motif for binding DNA in the unphosphorylated state. Phosphorylation increases DNA-binding affinity by ~10-fold, presumably through dimerization of the receiver domain that brings two effector domains next to each other for binding to DNA inverted repeats. DosR from *M. tuberculosis* forms a dimer with a unique structure [45]. Its receiver domain has a $(\beta\alpha)_4$ fold instead of the $(\beta\alpha)_5$ fold observed for all other RRs (Figure 13b), even though the conserved residues in the $\beta 5$ to $\alpha 5$ region are present in the DosR sequence. The sequence for the canonical $\beta 5$ to $\alpha 5$ region forms a long helix, pairing with a helix of linker residues to form an antiparallel coiled-coil (Figure 13b). These two helices form an antiparallel four-helix bundle with the same helices from another subunit, accounting for a major part of the dimer interface. Helices $\alpha 9$ and $\alpha 10$ have extensive interactions with the receiver domain, locking the structure in an inactive conformation. Activation by phosphorylation presumably causes significant structural

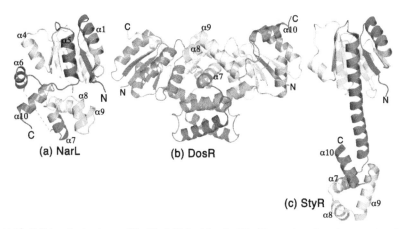

Figure 13. Full-length structures of the NarL/FixJ subfamily RRs. The receiver domains are colored in a rainbow spectrum from N-terminus in blue to the linker in red; the effector domains are separately colored in another rainbow spectrum from blue (α7) to red (α10). (a) NarL (PDB code 1RNL) has a tight packing between domains that blocks sequence recognition helix α9. Part of the linker that is disordered is depicted as a dashed line. (b) DosR (3C3W) forms a dimer, and one of the subunit is colored in gray for the receiver domain and the linker, and in pale green for the effector domain. (c) StyR (1ZN2) has a long helix α5 continuous into the linker that separates the two domains apart.

rearrangements that allows the receiver domain to form a dimer of (βα)₅ fold and release the effector domain to bind DNA inverted repeats [46].

OmpR/PhoB subfamily RRs have been extensively studied in recent years. In addition to many isolated receiver and effector domain structures, there are six full-length RR structures available, DrrB [47] and DrrD [48] from *T. maritima,* and RegX3 [49], PrrA [35], MtrA [50], and PhoP [34] from *M. tuberculosis.* All structures are in the unphosphorylated forms. The interdomain interfaces are different among these structures. RegX3 forms a domain-swapped dimer in the crystal structure, exchanging the α4-β5-α5 unit between subunits. It is not clear if this domain-swapped dimer is physiologically relevant. However, consistent with the structures of the isolated receiver domains, the α4-β5-α5 unit has high mobility among the structures of full-length RRs, indicating the importance of the α4-β5-α5 face in regulating RR activities. All structures with interactions between the receiver domain and effector domain involve the α4-β5-α5 face in the interdomain interface (Figure 14).

The structures of DrrD and DrrB both reveal monomers in crystal and have interactions between domains that do not preclude DNA binding (Figure 14). The interdomain interactions are relatively weak, suggesting that in solution the domains could be free and thus phosphorylation activation mechanism is likely through dimerization of the N-terminal receiver domain. The structures of PrrA and MtrA show more extensive interdomain interactions that involve the α4-β5-α5 face and block the recognition helix. However, the interfaces are polar in nature, and the open conformation is likely to exist in solution for both proteins. All these structural data suggest that phosphorylation is likely to shift the equilibrium toward dimerization as a mechanism to activate the effector domain.

The structure of PhoP from *M. tuberculosis* supports this mechanism of activation through phosphorylation-induced dimerization of the receiver domain [34]. PhoP forms a dimer through the α4-β5-α5 face of the receiver domain (Figure 14). The effector domain is merely tethered to the receiver domain through a flexible linker, which is disordered in the crystal structure. This open conformation of unphosphorylated PhoP structure is consistent with observations that unphosphorylated PhoP binds DNA [51]. Because there is no direct interaction between the receiver domain and the effector domain, it is likely that phosphorylation regulates the DNA-binding activity by shifting the equilibrium toward dimer formation of the receiver domain [33, 44]. Although the PhoP dimer interface involves the α4-β5-α5 face, the dimer packing is less compact than the activated receiver dimers. This is because the switch residues are in the non-activated conformation, and hence the loops at the dimer interface are different in conformation from the activated structures [34]. By shifting the equilibrium toward dimer formation of the receiver domains, phosphorylation brings two DNA-binding domains in close proximity; the flexible domain linker allows the DNA-binding domains to bind two direct repeats of DNA sequence while the receiver domains adopt a symmetric dimer. The OmpR/PhoB subfamily RRs recognize DNA direct-repeat sequences as revealed in the PhoB effector domain structure in complex with DNA (Figure 12a). However, the isolated DNA-binding domain of PhoP is a monomer in solution and does not form any dimers in the crystal structure [52]. Recognition of DNA direct repeats can only be realized through dimerization of the receiver domain. Bringing two DNA-binding domains in close proximity by dimerization of the receiver domains can also increase DNA-binding affinity.

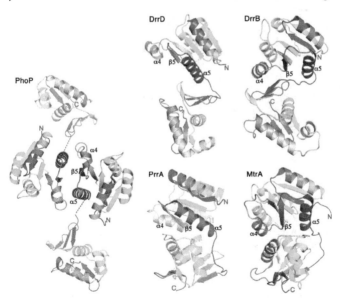

Figure 14. Full-length structures of the OmpR/PhoB subfamily RRs. Each domain is individually colored from N- to C-terminus in a rainbow spectrum. PhoP, PrrA, and MtrA are from *M. tuberculosis*; DrrD and DrrB are from *T. maritima*. The PDB codes are: PhoP, 3R0J; DrrD, 1KGS; DrrB, 1P2F; PrrA, 1YS6; MtrA, 2GWR.

4. Interactions between histidine kinases and response regulators

To complete the signaling events of two-component systems, histidine kinases, when activated, must associate with their cognate response regulators and transfer the phosphoryl group from the phosphorylated histidine to a conserved aspartate residue of RRs. Evidence suggests that HKs are also involved in dephosphorylation of RRs to turn off the signal cascade. The interactions between HK and RR must be specific because there is little cross-talk among TCS proteins. Yet, these interactions must be dynamic, as both proteins must adopt various conformations throughout the steps of the signaling event.

Several crystal structures of HK in complex with RR have been reported that offer insights into how HK and RR interact. A structure of the complex between HK853 and RR468 from *T. maritima* shows the two proteins engaged in a conformation for the phosphatase reaction [28]. The structure contains a dimer of the HK853 entire cytoplasmic portion and two RR468 molecules (Figure 15).

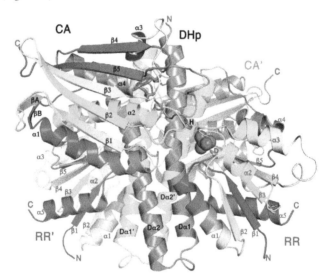

Figure 15. Structure of the HK853-RR468 complex (PDB code 3DGE). The DHp domain is colored in orange for one subunit and in green for the other. One CA domain (front) is colored in a rainbow spectrum; the other CA (back) is in sand color. Two RR molecules each individually colored in a rainbow spectrum. Secondary structural elements of HK are labeled in black, and those of RR are in purple. A sulfate between the essential histidine of HK and aspartate of RR is shown in a space-filling model.

The RR binds to the DHp domain on the helical hairpin below the phosphorylation site histidine. Helix α1 of RR interacts with both helices of the helical hairpin through hydrophobic side chains to form a six-helix bundle. Each α1 interacts only with helices of the same subunit of DHp. Loop β5-α5 also makes contacts with the DHp domain. In the crystal structure, there is a sulfate ion bound between the phosphoacceptor aspartate of RR and the phosphorylation site histidine of DHp. Binding of the sulfate mimics phosphorylation and allows the switch residues of RR to adopt the active conformation. Consequently, both loops β3-α3 and β4-α4 are also in the active conformation. These two loops interact with the ATP-lid and loop α2-β2 of the CA domain. The interactions between HK853 and RR468 would preclude dimer formation of RR through the α4-β5-α5 face, despite that the RR is in an active conformation. RR468 is a single domain protein. The structure of activated RR468 with BeF$_3^-$ shows changes in conformation of switch residues and loops β3-α3 and β4-α4, but it is a monomer [28]. This is different from OmpR/PhoB subfamily RRs, whose activated form promotes dimerization, or NarL/FixJ family RRs, whose activated form releases the effector domain for binding DNA and might also promote dimerization. Despite these differences, binding interactions of α1 of RR with the helical hairpin of DHp is likely conserved among TCS proteins. The structure of HK853-RR468 complex is in a phosphatase-competent state because the RR is phosphorylated and the ATP-lid blocks access of the γ-phosphate of ATP by the histidine side chain. The structure indicates that the phosphorylation site histidine could play a role in the phosphatase activity by orienting a water molecule as a nucleophile.

A complex structure of ThkA/TrrA, also from *T. maritima*, has a similar binding site on HK for RR [53], and the structure also represents a phosphatase-competent state because the CA domain is locked in a position away from the phosphorylation site histidine. Similar to RR468, TrrA is also a single domain RR. The resolution of the complex structure is relatively low at 3.8 Å, and high-resolution structures of individual domains were fit into the low-resolution electron density map of the ThkA-TrrA complex. TrrA has weak electron density, indicating its high mobility in the structure. The complex puts the phosphoacceptor aspartate of TrrA facing the phosphorylation site histidine of ThkA. However, TrrA has weaker interactions with DHp. Most of interactions involve the loops around the phosphoacceptor and the N-terminal end of α1 of TrrA with Dα1 of ThkA. Unlike that of RR468, helix α1 of TrrA does not have exposed hydrophobic side chains to have coiled-coil interactions with the four-helix bundle of ThkA. The binding position of RR on DHp is similar to that of the HK853-RR468 complex, suggesting that they share a common mechanism for the phosphotransfer reaction.

A structure of the SpoOF-SpoOB complex is related to HK-RR complexes in binding interactions and the mechanism of the catalyzed reactions. SpoOB is a phosphotransferase having a DHp domain similar to that of HKs, while SpoOF is a single domain RR [54, 55]. SpoOF is phosphorylated on its phosphoacceptor aspartate by sensor HK KinA. The phosphoryl group is then transferred to the phosphorylation site histidine on spoOB, which then transfers the phosphoryl group to the response regulator SpoOA. In the structure of the SpoOF-SpoOB complex, SpoOF binds to the four-helix bundle of SpoOB in a similar manner as the HK853-RR468 complex. Helix α1 of SpoOF packs against the helical hairpin region and makes up majority of the interface with DHp, confirming that helix α1 and the lower end of the DHp four-helix bundle as the binding interface between HK and RR.

5. Therapeutic potential against bacterial pathogens

Antibiotics resistance has been a major medical problem of modern medicine. Bacterial pathogens evolve mechanisms to become resistant to antibiotics, and strains resistant to multiple drugs or the last line of antibiotics, such as methicillin-resistant *S. aureus* and vancomycin-resistant *Enterococcus* have evolved and become a major challenge for modern medicine. Emergence of multi-drug resistant (MDR) and extremely drug resistant (XDR) tuberculosis (TB) has prompted the World Health Organization to declare TB as a global emergency.

Two component systems are major signaling proteins in bacteria. They are involved in every aspect of bacterial adaptation to their environments, including sensing the existence of antibiotics and regulating cellular responses to become drug resistant. Therefore, TCSs are potential drug targets for developing new antibiotics [56]. Unlike conventional antibiotics that directly target the proteins involved in essential cellular activities, drugs inhibiting TCSs target the upstream functions that regulate these essential proteins. Thus anti-TCS drugs work in a manner different from conventional drugs and are likely to be effective against drug-resistant bacterial pathogens. Because TCS proteins are absent in animals, drugs targeting TCSs can potentially have less toxicity.

Recent advancements in structural and functional characterization of TCSs have identified several potential targets for developing new antibiotics [56]. A PhoP-PhoQ system from *S. typhimurium* is a major virulence factor that is essential for this intracellular pathogen to establish infection. Inhibitors to the kinase activity of PhoQ were studied for their interactions with the protein and the inhibition mechanism by NMR and crystallography methods [57]. The structures can serve as a platform for rational design to improve the efficacy of the lead compounds. A PhoP-PhoR system in *M. tuberculosis* was found to be an important virulence factor [58]. PhoP has a point mutation in the avirulent H37Ra strain, and this mutation plays an important role in the loss of virulence of the avirulent strain [59]. Inhibitors targeting PhoP are likely to reduce the virulence of *M. tuberculosis*. With the availability of the PhoP structure [34], it is possible to do virtual screening or rational inhibitor design to search for compounds that will disrupt phosphorylation activation by interacting with the loops near the phosphoacceptor or with helix α1, or binding to the α4-β5-α5 face to prevent dimerization. Compounds inhibiting the effector domain binding DNA will also be efficient inhibitors. To develop inhibitors against PhoP-PhoR TCS, targeting the sensor domain of PhoR will be more specific and has the advantage that the compounds do not have to cross the cell membrane [58]. However, structural information of the PhoR protein is currently lacking.

Author details

Shuishu Wang

Department of Biochemistry and Molecular Biology, Uniformed Services University of the Health Sciences, Bethesda, Maryland, USA

Acknowledgement

The author thanks Dr. Chou-Zen Giam, Dr. Linda Miallau, and Dr. Issar Smith for helpful comments and revisions of the manuscript. This work was supported by a grant GM079185 from the National Institute of Health of the United States.

6. References

[1] Aravind L, Ponting CP. The cytoplasmic helical linker domain of receptor histidine kinase and methyl-accepting proteins is common to many prokaryotic signalling proteins. FEMS Microbiol Lett. 1999;176(1):111-6.

[2] Hefti MH, Francoijs KJ, de Vries SC, Dixon R, Vervoort J. The PAS fold. A redefinition of the PAS domain based upon structural prediction. Eur J Biochem. 2004;271(6):1198-208.

[3] Martinez SE, Beavo JA, Hol WG. GAF domains: two-billion-year-old molecular switches that bind cyclic nucleotides. Mol Interv. 2002;2(5):317-23.

[4] Wayne LG, Sohaskey CD. Nonreplicating persistence of mycobacterium tuberculosis. Annu Rev Microbiol. 2001;55:139-63.

[5] Cho HY, Cho HJ, Kim YM, Oh JI, Kang BS. Structural insight into the heme-based redox sensing by DosS from Mycobacterium tuberculosis. J Biol Chem. 2009;284(19):13057-67.

[6] Podust LM, Ioanoviciu A, Ortiz de Montellano PR. 2.3 A X-ray structure of the heme-bound GAF domain of sensory histidine kinase DosT of Mycobacterium tuberculosis. Biochemistry. 2008;47(47):12523-31.

[7] Miyatake H, Mukai M, Park SY, Adachi S, Tamura K, Nakamura H, et al. Sensory mechanism of oxygen sensor FixL from Rhizobium meliloti: crystallographic, mutagenesis and resonance Raman spectroscopic studies. J Mol Biol. 2000;301(2):415-31.

[8] Sevvana M, Vijayan V, Zweckstetter M, Reinelt S, Madden DR, Herbst-Irmer R, et al. A ligand-induced switch in the periplasmic domain of sensor histidine kinase CitA. J Mol Biol. 2008;377(2):512-23.

[9] Cheung J, Hendrickson WA. Sensor domains of two-component regulatory systems. Curr Opin Microbiol. 2010;13(2):116-23.

[10] Cheung J, Hendrickson WA. Structural analysis of ligand stimulation of the histidine kinase NarX. Structure. 2009;17(2):190-201.

[11] Moore JO, Hendrickson WA. Structural analysis of sensor domains from the TMAO-responsive histidine kinase receptor TorS. Structure. 2009;17(9):1195-204.

[12] Jing X, Jaw J, Robinson HH, Schubot FD. Crystal structure and oligomeric state of the RetS signaling kinase sensory domain. Proteins. 2010;78(7):1631-40.

[13] Airola MV, Watts KJ, Bilwes AM, Crane BR. Structure of concatenated HAMP domains provides a mechanism for signal transduction. Structure. 2010;18(4):436-48.

[14] Hulko M, Berndt F, Gruber M, Linder JU, Truffault V, Schultz A, et al. The HAMP domain structure implies helix rotation in transmembrane signaling. Cell. 2006;126(5):929-40.

[15] Appleman JA, Chen LL, Stewart V. Probing conservation of HAMP linker structure and signal transduction mechanism through analysis of hybrid sensor kinases. J Bacteriol. 2003;185(16):4872-82.

[16] Zhu Y, Inouye M. Analysis of the role of the EnvZ linker region in signal transduction using a chimeric Tar/EnvZ receptor protein, Tez1. J Biol Chem. 2003;278(25):22812-9.

[17] Ferris HU, Dunin-Horkawicz S, Mondejar LG, Hulko M, Hantke K, Martin J, et al. The mechanisms of HAMP-mediated signaling in transmembrane receptors. Structure. 2011;19(3):378-85.

[18] Ferris HU, Dunin-Horkawicz S, Hornig N, Hulko M, Martin J, Schultz JE, et al. Mechanism of regulation of receptor histidine kinases. Structure. 2012;20(1):56-66.

[19] Tomomori C, Tanaka T, Dutta R, Park H, Saha SK, Zhu Y, et al. Solution structure of the homodimeric core domain of Escherichia coli histidine kinase EnvZ. Nat Struct Biol. 1999;6(8):729-34.

[20] Varughese KI, Madhusudan, Zhou XZ, Whiteley JM, Hoch JA. Formation of a novel four-helix bundle and molecular recognition sites by dimerization of a response regulator phosphotransferase. Mol Cell. 1998;2(4):485-93.

[21] Marina A, Waldburger CD, Hendrickson WA. Structure of the entire cytoplasmic portion of a sensor histidine-kinase protein. Embo J. 2005;24(24):4247-59.

[22] Trajtenberg F, Grana M, Ruetalo N, Botti H, Buschiazzo A. Structural and enzymatic insights into the ATP binding and autophosphorylation mechanism of a sensor histidine kinase. J Biol Chem. 2010;285(32):24892-903.

[23] Marina A, Mott C, Auyzenberg A, Hendrickson WA, Waldburger CD. Structural and mutational analysis of the PhoQ histidine kinase catalytic domain. Insight into the reaction mechanism. J Biol Chem. 2001;276(44):41182-90.

[24] Albanesi D, Martin M, Trajtenberg F, Mansilla MC, Haouz A, Alzari PM, et al. Structural plasticity and catalysis regulation of a thermosensor histidine kinase. Proc Natl Acad Sci U S A. 2009;106(38):16185-90.

[25] Zhang Z, Hendrickson WA. Structural characterization of the predominant family of histidine kinase sensor domains. J Mol Biol. 2010;400(3):335-53.

[26] Gordeliy VI, Labahn J, Moukhametzianov R, Efremov R, Granzin J, Schlesinger R, et al. Molecular basis of transmembrane signalling by sensory rhodopsin II-transducer complex. Nature. 2002;419(6906):484-7.

[27] Goldberg SD, Clinthorne GD, Goulian M, DeGrado WF. Transmembrane polar interactions are required for signaling in the Escherichia coli sensor kinase PhoQ. Proc Natl Acad Sci U S A. 2010;107(18):8141-6.

[28] Casino P, Rubio V, Marina A. Structural insight into partner specificity and phosphoryl transfer in two-component signal transduction. Cell. 2009;139(2):325-36.

[29] Bick MJ, Lamour V, Rajashankar KR, Gordiyenko Y, Robinson CV, Darst SA. How to switch off a histidine kinase: crystal structure of Geobacillus stearothermophilus KinB with the inhibitor Sda. J Mol Biol. 2009;386(1):163-77.

[30] Galperin MY. Diversity of structure and function of response regulator output domains. Curr Opin Microbiol. 2010;13(2):150-9.

[31] Bourret RB. Receiver domain structure and function in response regulator proteins. Curr Opin Microbiol. 2010;13(2):142-9.

[32] Gao R, Stock AM. Molecular strategies for phosphorylation-mediated regulation of response regulator activity. Curr Opin Microbiol. 2010;13(2):160-7.

[33] Barbieri CM, Mack TR, Robinson VL, Miller MT, Stock AM. Regulation of response regulator autophosphorylation through interdomain contacts. J Biol Chem. 2010;285(42):32325-35.

[34] Menon S, Wang S. Structure of the response regulator PhoP from Mycobacterium tuberculosis reveals a dimer through the receiver domain. Biochemistry. 2011;50(26):5948-57.

[35] Nowak E, Panjikar S, Konarev P, Svergun DI, Tucker PA. The structural basis of signal transduction for the response regulator PrrA from Mycobacterium tuberculosis. J Biol Chem. 2006;281(14):9659-66.

[36] Bachhawat P, Stock AM. Crystal structures of the receiver domain of the response regulator PhoP from Escherichia coli in the absence and presence of the phosphoryl analog beryllofluoride. J Bacteriol. 2007;189(16):5987-95.

[37] Toro-Roman A, Wu T, Stock AM. A common dimerization interface in bacterial response regulators KdpE and TorR. Protein Sci. 2005;14(12):3077-88.

[38] Birck C, Mourey L, Gouet P, Fabry B, Schumacher J, Rousseau P, et al. Conformational changes induced by phosphorylation of the FixJ receiver domain. Structure. 1999;7(12):1505-15.

[39] Sidote DJ, Barbieri CM, Wu T, Stock AM. Structure of the Staphylococcus aureus AgrA LytTR domain bound to DNA reveals a beta fold with an unusual mode of binding. Structure. 2008;16(5):727-35.

[40] Blanco AG, Sola M, Gomis-Ruth FX, Coll M. Tandem DNA recognition by PhoB, a two-component signal transduction transcriptional activator. Structure. 2002;10(5):701-13.

[41] Maris AE, Kaczor-Grzeskowiak M, Ma Z, Kopka ML, Gunsalus RP, Dickerson RE. Primary and secondary modes of DNA recognition by the NarL two-component response regulator. Biochemistry. 2005;44(44):14538-52.

[42] Maris AE, Sawaya MR, Kaczor-Grzeskowiak M, Jarvis MR, Bearson SM, Kopka ML, et al. Dimerization allows DNA target site recognition by the NarL response regulator. Nat Struct Biol. 2002;9(10):771-8.

[43] Baikalov I, Schroder I, Kaczor-Grzeskowiak M, Grzeskowiak K, Gunsalus RP, Dickerson RE. Structure of the Escherichia coli response regulator NarL. Biochemistry. 1996;35(34):11053-61.

[44] Milani M, Leoni L, Rampioni G, Zennaro E, Ascenzi P, Bolognesi M. An active-like structure in the unphosphorylated StyR response regulator suggests a phosphorylation-dependent allosteric activation mechanism. Structure. 2005;13(9):1289-97.

[45] Wisedchaisri G, Wu M, Sherman DR, Hol WG. Crystal structures of the response regulator DosR from Mycobacterium tuberculosis suggest a helix rearrangement mechanism for phosphorylation activation. J Mol Biol. 2008;378(1):227-42.

[46] Wisedchaisri G, Wu M, Rice AE, Roberts DM, Sherman DR, Hol WG. Structures of Mycobacterium tuberculosis DosR and DosR-DNA complex involved in gene activation during adaptation to hypoxic latency. J Mol Biol. 2005;354(3):630-41.

[47] Robinson VL, Wu T, Stock AM. Structural analysis of the domain interface in DrrB, a response regulator of the OmpR/PhoB subfamily. J Bacteriol. 2003;185(14):4186-94.

[48] Buckler DR, Zhou Y, Stock AM. Evidence of intradomain and interdomain flexibility in an OmpR/PhoB homolog from Thermotoga maritima. Structure (Camb). 2002;10(2):153-64.

[49] King-Scott J, Nowak E, Mylonas E, Panjikar S, Roessle M, Svergun DI, et al. The structure of a full-length response regulator from Mycobacterium tuberculosis in a stabilized three-dimensional domain-swapped, activated state. J Biol Chem. 2007;282(52):37717-29.

[50] Friedland N, Mack TR, Yu M, Hung LW, Terwilliger TC, Waldo GS, et al. Domain orientation in the inactive response regulator Mycobacterium tuberculosis MtrA provides a barrier to activation. Biochemistry. 2007;46(23):6733-43.

[51] Gupta S, Sinha A, Sarkar D. Transcriptional autoregulation by Mycobacterium tuberculosis PhoP involves recognition of novel direct repeat sequences in the regulatory region of the promoter. FEBS Lett. 2006;580(22):5328-38.

[52] Wang S, Engohang-Ndong J, Smith I. Structure of the DNA-binding domain of the response regulator PhoP from Mycobacterium tuberculosis. Biochemistry. 2007;46(51):14751-61.

[53] Yamada S, Sugimoto H, Kobayashi M, Ohno A, Nakamura H, Shiro Y. Structure of PAS-linked histidine kinase and the response regulator complex. Structure. 2009;17(10):1333-44.

[54] Zapf J, Sen U, Madhusudan, Hoch JA, Varughese KI. A transient interaction between two phosphorelay proteins trapped in a crystal lattice reveals the mechanism of molecular recognition and phosphotransfer in signal transduction. Structure. 2000;8(8):851-62.

[55] Varughese KI, Tsigelny I, Zhao H. The crystal structure of beryllofluoride Spo0F in complex with the phosphotransferase Spo0B represents a phosphotransfer pretransition state. J Bacteriol. 2006;188(13):4970-7.

[56] Gotoh Y, Eguchi Y, Watanabe T, Okamoto S, Doi A, Utsumi R. Two-component signal transduction as potential drug targets in pathogenic bacteria. Curr Opin Microbiol. 2010;13(2):232-9.

[57] Guarnieri MT, Zhang L, Shen J, Zhao R. The Hsp90 inhibitor radicicol interacts with the ATP-binding pocket of bacterial sensor kinase PhoQ. J Mol Biol. 2008;379(1):82-93.

[58] Ryndak M, Wang S, Smith I. PhoP, a key player in Mycobacterium tuberculosis virulence. Trends Microbiol. 2008;16(11):528-34.

[59] Frigui W, Bottai D, Majlessi L, Monot M, Josselin E, Brodin P, et al. Control of M. tuberculosis ESAT-6 secretion and specific T cell recognition by PhoP. PLoS pathogens. 2008;4(2):e33.

Permissions

The contributors of this book come from diverse backgrounds, making this book a truly international effort. This book will bring forth new frontiers with its revolutionizing research information and detailed analysis of the nascent developments around the world.

We would like to thank Cai Huang, Ph.D., for lending his expertise to make the book truly unique. He has played a crucial role in the development of this book. Without his invaluable contribution this book wouldn't have been possible. He has made vital efforts to compile up to date information on the varied aspects of this subject to make this book a valuable addition to the collection of many professionals and students.

This book was conceptualized with the vision of imparting up-to-date information and advanced data in this field. To ensure the same, a matchless editorial board was set up. Every individual on the board went through rigorous rounds of assessment to prove their worth. After which they invested a large part of their time researching and compiling the most relevant data for our readers. Conferences and sessions were held from time to time between the editorial board and the contributing authors to present the data in the most comprehensible form. The editorial team has worked tirelessly to provide valuable and valid information to help people across the globe.

Every chapter published in this book has been scrutinized by our experts. Their significance has been extensively debated. The topics covered herein carry significant findings which will fuel the growth of the discipline. They may even be implemented as practical applications or may be referred to as a beginning point for another development. Chapters in this book were first published by InTech; hereby published with permission under the Creative Commons Attribution License or equivalent.

The editorial board has been involved in producing this book since its inception. They have spent rigorous hours researching and exploring the diverse topics which have resulted in the successful publishing of this book. They have passed on their knowledge of decades through this book. To expedite this challenging task, the publisher supported the team at every step. A small team of assistant editors was also appointed to further simplify the editing procedure and attain best results for the readers.

Our editorial team has been hand-picked from every corner of the world. Their multi-ethnicity adds dynamic inputs to the discussions which result in innovative outcomes. These outcomes are then further discussed with the researchers and contributors who give their valuable feedback and opinion regarding the same. The feedback is then collaborated with the researches and they are edited in a comprehensive manner to aid the understanding of the subject.

Apart from the editorial board, the designing team has also invested a significant amount of their time in understanding the subject and creating the most relevant covers. They scrutinized every image to scout for the most suitable representation of the subject and create an appropriate cover for the book.

The publishing team has been involved in this book since its early stages. They were actively engaged in every process, be it collecting the data, connecting with the contributors or procuring relevant information. The team has been an ardent support to the editorial, designing and production team. Their endless efforts to recruit the best for this project, has resulted in the accomplishment of this book. They are a veteran in the field of academics and their pool of knowledge is as vast as their experience in printing. Their expertise and guidance has proved useful at every step. Their uncompromising quality standards have made this book an exceptional effort. Their encouragement from time to time has been an inspiration for everyone.

The publisher and the editorial board hope that this book will prove to be a valuable piece of knowledge for researchers, students, practitioners and scholars across the globe.

List of Contributors

Dmytro Pavlov and John G. Mielke
School of Public Health and Health Systems, University of Waterloo, Waterloo, Canada

Masataka Oda, Masahiro Nagahama, Keiko Kobayashi and Jun Sakurai
Faculty of Pharmaceutical Sciences, Tokushima Bunri University, Japan

Nina Kurrle, Bincy John, Melanie Meister and Ritva Tikkanen
Institute of Biochemistry, Medical Faculty, University of Giessen, Giessen, Germany

Yamuna D. Gandaharen and Sydney Welt
Welt Bio-Molecular Pharmaceutical, LLC, Molecular Research, Armonk, NY, U.S.A.

Rachel S. Welt
Department of Biological Sciences, Fordham University, NYC, NY, USA

David Kostyal
ARD Laboratories, Biologic Research, Akron, OH, USA

C. Frazer
Department of Biomedical and Molecular Science, Queen's University, Kingston, Ontario, K7L 3N6, Canada

P.G. Young
Department of Biology, Queen's University, Kingston, Ontario, Canada

Roberta Fraschini, Erica Raspelli and Corinne Cassani
Università degli Studi di Milano-Bicocca, Dipartimento di Biotecnologie e Bioscienze, Milano, Italy

Shuishu Wang
Department of Biochemistry and Molecular Biology, Uniformed Services University of the Health Sciences, Bethesda, Maryland, USA

Printed in the USA
CPSIA information can be obtained
at www.ICGtesting.com
JSHW011407221024
72173JS00003B/451